Pump Wisdom

Pump Wisdom

Problem Solving for Operators and Specialists

by
Heinz P. Bloch

AIChE

WILEY

A JOHN WILEY & SONS, INC., PUBLICATION

Cover based on ideas by Dennis Bowman, PE, Alfred Conhagen, Inc., Houston, Texas. Page layout and formatting by Joy M. Peterson, Asheville, North Carolina.

For general information on our other products and services or for technical support, please contact our Customer Care Department within the United States at (800) 762-2974, outside the United States at (317) 572-3993 or fax (317) 572-4002.

Wiley also publishes its books in a variety of electronic formats. Some content that appears in print may not be available in electronic formats. For more information about Wiley products, visit our web site at www.wiley.com.

Library of Congress Cataloging-in-Publication Data:

Bloch, Heinz P., 1933–
  Pump wisdom : problem solving for operators and specialists / Heinz P. Bloch.
      p. cm.
  Includes index.
    ISBN 978-1-118-04123-9 (hardback)
  1. Pumping machinery. I. Title.
  TJ900.B6483 2011
  621.6'9—dc22   2010050397

Printed in the United States of America.

10 9 8 7 6 5 4 3 2 1

To Ken and the fellow professionals
who appreciate the importance of
avoiding pump failures

Table of Contents

Spiral wound and kammprofile gaskets
Pipe, hydraulic tubing, or flexible connections?
Gusseting
Concentric versus eccentric reducers
Vibration problems in piping
What we have learned on piping and gasketing topics
References

Bearing selection overview and windage as a design problem
Radial versus axial (thrust) bearings
Oil levels, multiple bearings, and different bearing orientations
Upgrading and retrofit opportunities
Bearing cages
Bearing preload and clearance effects
Bearing dimensions and mounting tolerances
What we have learned
References

Lubricant level and oil application
Issues with oil rings
Pressure and temperature balance in bearing housings
Cooling not needed on pumps with rolling element bearings
Oil delivery by constant level lubricators
Black oil
Lube application as oil mist (oil fog)
Desiccant breathers and expansion chambers
What we have learned
References

Lubricant viscosities
When and why high film strength synthetic lubricants are used
Lubricants for oil mist systems
What we have learned
References

Avoid solids contamination
Avoid questionable storage and transfer practices
Periodic audits
What we have learned
References

Vibration and its effect on bearing life
Monitoring methods differ
Vibration acceptance limits
Causes of excessive vibration
Rotor balancing
What we have learned
References

Driver selection
Coupling selection and installation
Installation and removal
Alignment and quality criteria
Consequences of misalignment
Thermal rise and predefinition of growth
What we have learned
References

Low incremental cost of better pumps
Pump pedestals and bearing housings should not be water-cooled
Summary of bearing-related issues
Constant level lubricators
Bearing housing protector seals ("bearing isolators")
Motor lubrication summary
Mechanical seal issues
Hydraulic issues
Impeller hydraulics
Mechanical improvement or upgrade options
Process pump repair dimensions

Preface

Pump users have access to hundreds of books and many thousands of articles dealing with pump subjects. So, one might ask, why do we need *this* text? I believe we need it because an unacceptably large number of process pumps fail catastrophically every year. An estimated 95% of these are repeat failures, and most of them are costly, dangerous, or both. I wanted to explain some elusive failure causes and clearly map out permanent remedial action. My intent was to steer clear of the usual consultant-conceived generalities and give you tangible, factual, and well-defined information throughout.

As any close review of what has been offered in the past will uncover, many texts were written primarily to benefit one particular job function, ranging from pump operators to pump designers. Some books contain a hidden bias, or they appeal to a very narrow spectrum of readers; others are perhaps influenced by a particular pump manufacturer's agenda. Give *this* text a chance; you'll see that it is different. I gave it the title *Pump Wisdom* because wisdom is defined as applied knowledge. If you concur with this very meaningful definition you will be ready for a serious challenge. That challenge is to practice wisdom by acting on the knowledge this text conveys.

Although I had written or co-authored other books and dozens of articles on pump reliability improvement, some important material is too widely dispersed to be readily accessible. Moreover, some important material has never been published before. I therefore set out to assemble, rework, condense, and explain the most valuable points in a text aimed at wide distribution. It was to become a text with, I hope, permanence and "staying power." To thus satisfy the scope and intent of this book, I endeavored to keep theoretical explanations to a reasonable minimum and to limit the narrative to 200 pages. Putting it another way, I wanted to squeeze into these 200 pages material and topics that will greatly enhance both pump safety and reliability. All that is needed is the reader's solid determination to pay close attention and to follow up diligently.

Please realize that in years past, many pump manufacturers have primarily concentrated their design and improvement focus on the hydraulic end. Indeed, over time and in the decades since 1960, much advancement has been made in the metallurgical and power-efficiency-related performance of the hydraulic assembly. All the while, the mechanical assembly or drive end of process pumps was being treated with relative indifference. In essence, there was an imbalance between the attention given to pump hydraulics and pump mechanical issues.

Recognizing indifference as costly, this text tries to rectify some of these imbalances. My narrative and illustrations are intended to do justice to both the hydraulic assembly and the mechanical assembly of process pumps. That said, the book briefly lays out how pumps function and quickly moves to guidelines and details that must be considered by reliablity-focused readers. A number of risky omissions or shortcuts by pump designers, manufacturers, and users also are described.

I also want the reader to know that the real spark for this book came from AESSEAL, a worldwide manufacturer of sealing products. I have come to appreciate them as an entity guided by quality principles, with a workforce that pursues excellence at every level of the organization. Of the many business entities I have worked with over the past five decades, they represent the one closest to my own ideal. They practice what a reliability-focused company should be doing; I consider them an example of how a business should conduct itself. I also want to acknowledge them for rapidly providing artwork assistance whenever I asked.

Please take from me a good measure of encouragement: Make good use of this book. Read it and apply it. Today, and hopefully years from now, remember to consult this material. Doing so will acquaint you with pump failure avoidance and the more elusive aspects of preserving pump-related assets. And so, although you undoubtedly have more problems than you deserve, please keep in mind that you also deserve access to more solutions than you previously knew about or currently apply. Sound solutions are available, and they are here, right at your fingertips. Use them wisely; they will be cost-effective. The solutions you can discern from this text will have a positive effect on pump safety performance and asset preservation. They have worked at best-of-class companies and cannot possibly disappoint you.

Heinz P. Bloch, P.E.
Winter 2010/Spring 2011

Chapter 1

Principles of Centrifugal Process Pumps

Pumps, of course, are simple machines that lift, transfer, or otherwise move fluid from one place to another. They are usually configured to use the rotational (kinetic) energy from an impeller to impart motion to a fluid. The impeller is located on a shaft; together, shaft and impeller(s) make up the rotor. This rotor is surrounded by a casing; located in this casing (or pump case) are one or more stationary passageways that direct the fluid to a discharge nozzle. Impeller and casing are the main components of the *hydraulic assembly*; the region or envelope containing bearings and seals is called the mechanical assembly or power end (Figure 1.1).

Many process pumps are designed and constructed to facilitate field repair. On these so-called back pull-out pumps, shop maintenance can be performed, whereas the casing and its associated suction and discharge pip-

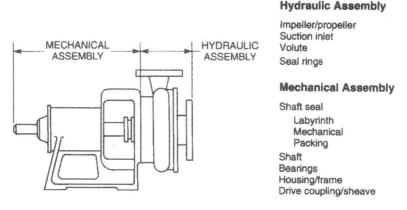

Hydraulic Assembly

Impeller/propeller
Suction inlet
Volute
Seal rings

Mechanical Assembly

Shaft seal
   Labyrinth
   Mechanical
   Packing
Shaft
Bearings
Housing/frame
Drive coupling/sheave

MECHANICAL ASSEMBLY — HYDRAULIC ASSEMBLY

Figure 1.1: Principal components of an elementary process pump (Ref. 1).

ing (Figure 1.2) are left undisturbed. Although operating in the hydraulic end, the impeller remains with the power end during removal from the field. The rotating impeller (Figure 1.3) is usually constructed with swept-back vanes and the fluid is accelerated from the rotating impeller to the stationary passages in the surrounding casing.

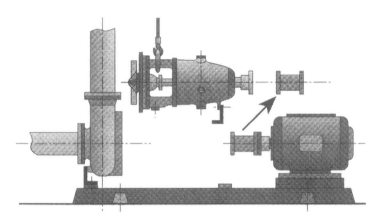

Figure 1.2: Typical process pump with suction flow entering horizontally and vertically oriented discharge pipe leaving the casing tangentially (Source: Ref. 2)

In this manner, kinetic energy is converted to potential energy and the fluid (also called pumpage) moves from the suction (lower) pressure side to the discharge (higher) pressure side of a pumping system. As the fluid leaves the impeller through the pump discharge, more fluid is drawn into the pump suction where, except for the region immediately adjacent, the pressure is lowest (Ref. 3).

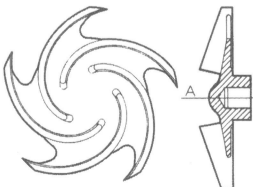

Figure 1.3: A semiopen impeller with five vanes. As shown, the impeller is configured for counterclockwise rotation about a centerline "A".

PUMP PERFORMANCE: HEAD AND FLOW

Pump performance is always described in terms of head H produced at a given flow capability Q and hydraulic efficiency η attained at any particular intersection of H and Q. Head is customarily plotted on the vertical scale or vertical axis (the left of the two y-axes) of Figure 1.4; it is expressed in feet (or meters). Hydraulic efficiency is often plotted on another vertical scale, at the right of the two vertical scales (i.e., the y-axis in this generalized plot).

Head is related to the difference between discharge pressure and suction pressure at the respective pump nozzles. Head is a simple concept, but this is where consideration of the impeller tip speed is important. The higher the shaft rpm and the larger the impeller diameter, the higher will be the impeller tip speed—actually its peripheral velocity.

The concept of head can be visualized by thinking of a vertical pipe bolted to the outlet (the discharge nozzle) of a pump. In this imaginary pipe, a column of fluid would rise to a height "H." If the vertical pipe would be attached to the discharge nozzle of a pump with *higher* impeller tip speed, then the fluid would rise to a greater height "H+." It is important to note that the height of a column of liquid, H or H+, is a function only of the impeller tip speed. The specific gravity of the liquid affects power demand but does not influence either H or H+. However, the resulting discharge pressure *does* depend on the liquid density (specific gravity or Sp.G.). For water (with an Sp.G. of 1.0), an H of 2.31 feet equals 1 psi (pound-per-square-inch), whereas for alcohol, which might have a Sp.G. of 0.5, a column height or head H of 4.62 feet equals 1 psi. So, if a certain fluid had an Sp.G. of 1.28, a then column height (head H) of 2.31/1.28 = 1.8 feet would equal a pressure of 1 psi.

For reasons of material strength and reasonably priced metallurgy, one usually limits the head per stage to about 700 feet. This is a fairly important rule-of-thumb limit to remember. When too many similar rule-of-thumb *limits* combine, one cannot expect pump reliability to be at its highest. As an example, say a particular impeller-to-shaft fit is to have 0.0002- to 0.0015-inch clearance on average size impeller hubs. With a clearance fit at the high limit of 0.0015 inches, one might anticipate a somewhat greater failure risk if bearing fits, coupling fits, and seal fits were all at their uppermost limits.

On Figure 1.4, the point of zero flow (where the curve intersects the y-axis) is called the shut-off point. The point at which operating efficiency is at a peak is called the best efficiency point, or BEP. Head rise from BEP to shut-off is often chosen around 10-15% of differential head. This choice makes it

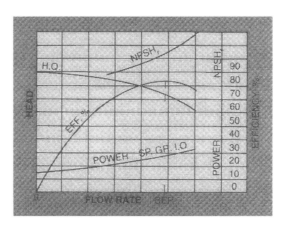

Figure 1.4: Typical "H-Q" performance curves are sloped as shown here. The BEP is marked with a small triangle; power and other parameters are often displayed on the same plot.

easy to modulate pump flow by adjusting the control valve open area based on monitoring pressure. Pumps "operate on their curves" and knowledge of what pressure relates to what flow allows technicians to program control loops.

The generalized depictions in Figure 1.4 also contain a curve labeled NPSHr, which stands for Net Positive Suction Head required. This is the head of liquid that must exist at the edge of the inlet vanes of an impeller to allow liquid transport without causing undue vaporization. It is a function of impeller geometry and size and is determined by factory testing. NPSHr can range from a few feet to a three-digit number. At all times, the head of liquid *available* at the impeller inlet (NPSHa) must exceed the required NPSHr.

Operation at Zero Flow

The rate of flow through a pump is labeled Q (gpm) and is plotted on the horizontal axis (the x-scale). Note that, for a given speed and for every value of head H we read off on the y-axis, there is a corresponding value of Q on the x-axis. This plotted relationship is expressed as "the pump is running on its curve." Pump H-Q curves are plotted to commence at zero flow and highest head. Process pumps need a continually rising curve inclination and a curve with a hump somewhere along its inclined line will not serve the reliability-focused user. Operation at zero flow is not allowed and, if over perhaps a minute's duration, could cause temperature increase and internal recirculation effects that might destroy most pumps.

But remember that this curve is valid only for this particular impeller pattern, geometry, size, and operation at the speed indicated by the manufacturer or entity that produced the curve. Curve steepness or inclination has to do with the number of vanes in that impeller; curve steepness is also affected by the angle each vane makes relative to the impeller hub. In general, curve shape is verified by physical testing at the manufacturer's facility. Once the entire pump is installed in the field, it can be retested periodically by the owner-purchaser for degradation and wear progression. Power draw may have been affected by seals and couplings that differ from the ones used on the manufacturer's test stand. Occasionally, high efficiencies are alluded to in the manufacturer's literature when bearing, seal, and coupling losses are not included in the vendor's test reports.

Impellers and Rotors

Regardless of flow classification, centrifugal pumps range in size from tiny pumps to big pumps. The tiny ones might be used in medical or laboratory applications; the extremely large pumps may move many thousands of liters or even gallons per second from flooded lowlands to the open sea.

All six impellers in Figure 1.5 are shown with a hub fastening the impeller to the shaft and each of the first five impellers is shown as a hub-and-disc version with an impeller cover. The cover (or "shroud") identifies the first five as "closed" impellers; recall that Figure 1.3 had depicted a semiopen impeller. Semiopen impellers are designed and fabricated without the cover. Finally, open impellers come with free-standing vanes welded to or integrally cast into the hub. Because the latter incorporate neither disc nor cover, they are often used in viscous or fibrous paper stock applications.

To function properly, a semiopen impeller must operate in close proximity to a casing internal surface, which is why axial adjustment features are needed with these impellers. Axial location is a bit less critical with closed impellers. Except on axial flow pumps, fluid exits the impeller in the radial direction. Radial and mixed flow pumps are either single or double suction designs; both will be shown later. Once the impellers are fastened to a shaft, the resulting assembly is called a rotor.

In radial and mixed flow pumps, the number of impellers following each other, typically called "stages," can range from one to as many as will make such multistage pumps practical and economical to manufacture. Horizontal shaft pumps with up to 12 stages are not uncommon. With longer rotors, it becomes more difficult to avoid operating with a high vibration resonance (so-called critical speed). Vertical shaft pumps have been designed with 48 or more stages. In vertical pumps, shaft support bushings are relatively lightly loaded; they are spaced so as to minimize vibration risk.

THE MEANING OF SPECIFIC SPEED

Pump impeller flow classifications and the general meaning of specific speed deserve additional discussion. Moving from left to right in Figure 1.5, the various impeller geometries reflect selections that start with high differential pressure capabilities and end with progressively lower differential pressure capabilities. Differential pressure is simply discharge pressure minus suction pressure.

Specific speed calculations are a function of several impeller parameters; the mathematical expression includes exponents and is found later in Figure 1.6. Staying with Figure 1.5 and again moving from left to right, we can reason that larger throughputs (flows) are more likely achieved by the configurations at the right, whereas larger pressure ratios (discharge pressure divided by suction pressure) are usually achieved by the impeller geometries closer to the left of the illustration.

Impellers toward the right are more efficient than those near the left, and pump designers use the parameter *specific speed* (N_s) to bracket pump hydraulic efficiency attainment and other expected attributes of a particular impeller configurations and size. Please be sure not to confuse a similar sounding parameter, *pump suction specific speed* (N_{ss} or N_{sss}), with the *specific speed* (N_s). For now, we are strictly addressing *specific speed* (N_s).

As an example, observe the customary use, whereby with N and Q—the typical given parameters that define centrifugal process pumps—one determines a pivot point. Next, with pivot point and head H, one can easily determine N_s. In Figure 1.5, N_s is somewhere between 500 and 15,000 on the U.S. scale. Whenever we find ourselves in that range, we know such a pump exists and we can even observe the general impeller shape. Keep in mind that thousands of impeller combination and geometries exist. Impel-

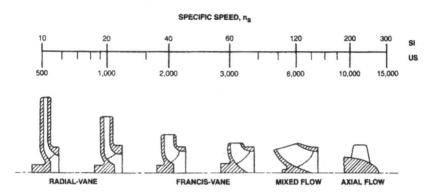

Figure 1.5: General flow classifications of process pump impellers.

lers with covers are the most prominent in hydrocarbon processes and an uneven number of impeller vanes is favored over even numbers of vanes for reasons of vibration suppression.

Pump specific speed, Figure 1.6, might be of primary interest to pump designers, but average users will also find it useful. On the lower right, the illustration gives the equation for N_s; it will be easy to see how N_s is related to the shaft speed N (rpm), flow Q (gpm or gallons/minute), and head H (expressed in feet). This mathematical expression also has two strange-looking exponents in it, but the N_s nomogram conveys more than meets the eye and can be helpful.

If now, we had the same or some other Q and would want to see what happens at some other speed, we would again draw a straight line to establish the pivot point. Drawing a line from whatever H is specified through the pivot point and to N_s, we would not like to select pumps with an N_s outside the rule-of-thumb range from 500 to 15,000. In another example, we might, after establishing the pivot point, wish to determine what happens if we select an impeller with the maximum head capability of 700 feet and draw a line through the pivot point. If the resulting N_s is too low, then we would try a higher speed N and see what happens.

Although there are always fringe applications in terms of size and flow rate, this book deals with centrifugal pumps in process plants. These pumps are related to the generic illustrations of Figures 1.1, 1.2, and others in this chapter. All would somewhat typically—but by no means exclusively—range from 3 to perhaps 300 hp (2-225 kW).

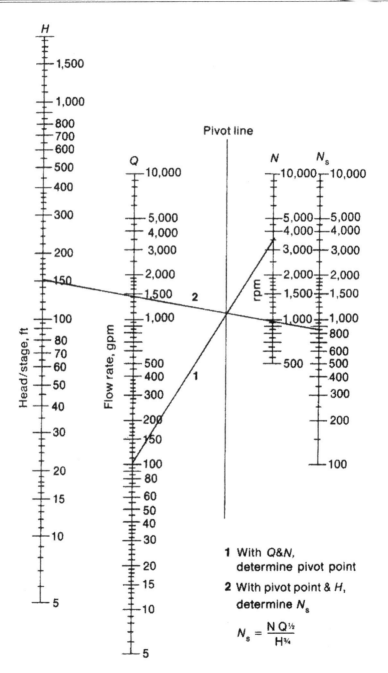

Figure 1.6: Pump specific speed nomogram allowing quick estimations.

PROCESS PUMP TYPES

The elementary process pumps illustrated in Figures 1.1 and 1.2 probably incorporate one of the radial vane impellers shown in Figure 1.5. If a certain differential pressure is to be achieved together with higher flows, such a pump is often designed with a double-flow impeller (Figure 1.7). One of the side benefits of double-flow impellers is good axial thrust equalization (axial balance). A small thrust bearing will often suffice; it is shown here in the left-bearing housing. Note that the two radial bearings are plain or sleeve-type. Certain sleeve bearings have relatively high speed capability.

If elevated pressures are needed, then several impellers are lined up in series on the same pump rotor. Of course, this would then turn the pump into a multistage model (Figure 1.8).

PROCESS PUMP MECHANICAL RESPONSE TO FLOW CHANGES

After the pumped fluid (also labeled pumpage, or flow) leaves at the impeller tip, it must be channeled into a stationary passageway that merges into the discharge nozzle. Many different types of passageway designs (single

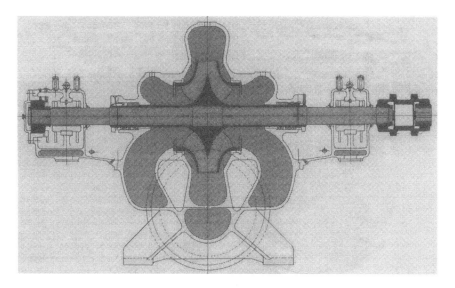

Figure 1.7: Double-flow impellers are used for higher flows and relatively equalized (balanced) axial hydrolic loads in both directions (Source: Ref. 4).

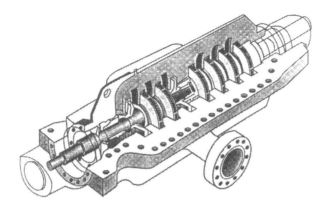

Figure 1.8: A multistage centrifugal process pump.

or multiple volutes, vaned diffusers, etc.) are available. Their respective geometry interacts with the flow and creates radial force action of different
magnitude around the periphery of an impeller (Figure 1.9). These forces
tend to deflect the pump shaft; they are greater at part flow than at full flow.

RECIRCULATION AND CAVITATION

Recirculation is a flow reversal near either the inlet or the discharge of a
centrifugal pump. This flow reversal produces cavitation erosion damage that
starts on the high-pressure side of an impeller vane and proceeds through the
metal to the low-pressure side (Ref. 5).

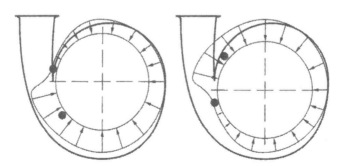

Figure 1.9: Direction and magnitude of fluid forces change at different flows (Ref.
6).

Pump-internal flow (called recirculation) can cause surging and cavitation even when the available NPSHa exceeds the manufacturer's published NPSHr by considerable margins. Also, extensive damage to the pressure side of impeller vanes has been observed in pumps operating at reduced flow rates. These are the obvious results of recirculation; however, more subtle symptoms and operational difficulties have been identified in pumps operating in the recirculation zone.

Symptoms of discharge recirculation include the following:
- Cavitation damage to the pressure side of the vane at the discharge
- Axial movement of the shaft, sometimes accompanied by damage to the thrust bearing
- Cracking or failure of the impeller shrouds at the discharge
- Shaft failure on the outboard end of double-suction and multistage pumps
- Cavitation damage to the casing tongue (see Chapter 11) or diffuser vanes

Symptoms of suction recirculation include the following:
- Cavitation damage to the pressure side of the vanes at the inlet
- Cavitation damage to the stationary vanes in the suction
- Random crackling noise in the suction; this contrasts with the steady crackling noise caused by inadequate net positive suction head
- Surging of the suction flow

Figure 1.10: Where and why impeller vanes get damaged (Ref. 5).

A quick-reference illustration was provided by Warren Fraser (Figure 1.10). We should note that recirculation and the attendant failure risks are low in pumps delivering 2,500 gpm or less at heads up to 150 feet. In those pumps, energy levels may not be high enough even if the pump operates in the recirculation zone. As a general rule for such pumps, minimum flow can be set at 50% of recirculation flow for continuous operation and at 25% of recirculation flow for intermittent operation (Ref. 7).

THE IMPORTANCE OF SUCTION SPECIFIC SPEED

Note that pump suction specific speed (N_{ss} or N_{sss}) differs from the pump specific speed N_s discussed earlier. For installations delivering more than 2,500 gpm and with suction specific speeds more than 9,000, greater care is needed. Suction specific speed (N_{sss} or N_{ss}) is calculated by the straightforward mathematical expression:

$$N_{ss} = \frac{(r/min)[(gal/min)/eye]^{1/2}}{(NPSH_r)^{3/4}}$$

wherein both the flow rate and the NPSHr pertain to conditions published by the manufacturer. In each case, these conditions (flow in gpm) and NPSHr are observed on the maximum available impeller diameter for that particular pump.

The higher the design suction specific speed, the closer the point of suction recirculation to what is commonly described as rated capacity. Similarly, the closer the discharge recirculation capacity is to rated capacity, the higher the efficiency. Pump system designers are tempted to aim for the highest possible efficiency and suction specific speed. However, such designs might result in systems with either a limited pump operating range or, if operated inside the recirculation range, a disappointing reliability and frequent failures.

Although more precise calculations are available, trend curves of probable NPSHr for minimum recirculation and zero cavitation erosion in water (Figure 1.11), are sufficiently accurate to warrant our attention (Ref. 8). The NPSHr needed for zero damage to impellers and other pump components may be many times that published in the manufacturer's literature. The manufacturers' NPSHr plot (lowermost curve in Figure 1.11) is based on observing a 3% drop in discharge head or pressure; at Q = 100% we note

TREND OF PROBABLE NPSH$_R$ FOR ZERO CAVITATION-EROSION

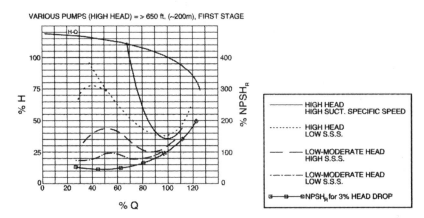

Figure 1.11: **Pump manufacturers usually plot only the NPSHR trend associated with the lowermost curve. At that time a head drop or pressure fluctuation of 3% exists and cavitation damage is often experienced (Ref. 7).**

NPSHr = 100% of the manufacturer's claim. Unfortunately, whenever this 3% fluctuation occurs, a measure of damage may already be in progress. Assume the true NPSHr is as shown in Figure 1.11 and aim to provide an NPSHa in excess of this true NPSHr.

In Ref. 8 Irving Taylor compiled his general observations and alerted us to this fact. He cautioned against considering his curves totally accurate and mentioned the demarcation line between low and high suction specific speeds somewhere between 8,000 and 12,000. Many data points taken after 1980 point to 8,500 or 9,000 as numbers of concern. If pumps with N_{ss} numbers higher than 9,000 are being operated at flows much higher or lower than BEP, then their life expectancy or repair-free operating time will be reduced.

In the decades since Taylor made his first presentation, controlled testing has been done in many industrialized countries. The various findings have been reduced to relatively accurate calculations that were later published by the Hydraulic Institute (HI) (Ref. 9). Relevant summaries can also be found in Ref. 10. Calculations based on Refs. 9 and 10 determine minimum allowable flow as a percentage of BEP.

Note, again, that recirculation differs from cavitation, a term that essentially describes vapor bubbles that collapse. Cavitation damage is often caused by low net positive suction head available (NPSHa). Such cavitation-related damage starts on the low-pressure side and proceeds to the high-pressure side. An impeller requires a certain net positive suction head; this NPSHr is simply the pressure needed at the impeller inlet (or eye) for relatively vapor-free flow.

What We Have Learned

Understanding the concepts of N_s and N_{ss} will assist in specifying better pumps. In addition to fluid properties, pump life is influenced by throughput. Just as an automobile transmission is designed to work best at particular speeds or optimum gear ratios, pumps have desirable flow ranges. Deviating from optimum flow will influence failure risk and life expectancy.

References
1. SKF USA, Inc., Kulpsville, PA; Publication 100-955, "Bearings in Centrifugal Pumps," Version 4/2008; excerpted or adapted by permission of the copyright holder.
2. Emile Egger & Cie., Salt Lake City, UT, and Cressier, NE (Switzerland).
3. ITT/Goulds Pump Corporation, Seneca Falls, NY; Installation and Maintenance Manual for Model 3196 ANSI Pump (1990).
4. Mitsubishi Heavy Industries, Ltd., Tokyo, Japan, and New York, NY; Publication HD30-04060).
5. Fraser, Warren H.; "Avoiding recirculation in centrifugal pumps," *Machine Design*, June 10, 1982.
6. *World Pumps*, February 2010, p. 19.
7. Ingram, James H.; "Pump reliability—where do you start," presented at ASME Petroleum Mechanical Engineering Workshop and Conference, Dallas, TX, September 13-15, 1981.
8. Taylor, Irving; "The most persistent pump-application problems for petroleum and power engineers," ASME Publication 77-Pet-5, Energy Technology Conference and Exhibit, Houston, TX, September 18-22, 1977.
9. "Allowable Operating Region," ANSI/HI9.6.3-1997, Hydraulic Institute, Parsippany, NJ.
10. Bloch, Heinz P., and Alan R. Budris; *Pump User's Handbook*. 3rd Edition (2010), The Fairmont Press, Lilburn, GA (ISBN 0-88173-627-9).

Chapter 2

Pump Selection and Industry Standards

Anybody can buy a cheap pump, but you want to buy a better pump. The term "better pumps" describes fluid movers that are well designed beyond just hydraulic efficiency and modern metallurgy. Better pumps are ones that avoid risk areas in the mechanical portion commonly called the drive-end. This is the part of process pumps that has been neglected most often and where cost-cutting should cause the greatest concern.

Deviations from best available technology increase the failure risk. As three or four or more deviations combine, a failure is likely to occur. An analogy could be drawn from an incident involving two automobiles, with one driving behind the other. When the trailing vehicle traveled at (a) an excessive speed, with (b) worn tires, on (c) a wet road, and (d) followed the leading car too closely, a rear-end collision resulted. If there had just been any three of the four violations, then the event might be recalled as one of the many "near miss" incidents. If there had been any two of the four, then it would serve no purpose to tell the story in the first place.

Too much cost-cutting by pump manufacturers and purchasers will negatively affect the drive-ends of process pumps. Flawed drive-end components are therefore among the main contributors to elusive repeat failures that often plague pumps—essentially simple machines. Drive-end flaws deserve to be addressed with urgency and this short chapter will introduce the reader to more details that follow in later chapters.

WHY INSIST ON BETTER PUMPS?

Well-informed reliability professionals will be reluctant to accept pumps that incorporate the drive-end shown in Figure 2.1. The short overview of reasons is that reliability-focused professionals take seriously their obligation to consider the actual, *lifetime-related* and not just *short-term*, cost of ownership. They have learned long ago that price is what one pays, and value is what one gets.

Anyway, although at first glance the reader might see nothing wrong, Figure 2.1 contains clues as to why many pumps fail relatively frequently and sometimes randomly. It shows areas of vulnerability that must be recognized and eliminated. The best time to eliminate flaws is in the specification process. The following important vulnerabilities, deviations from best available technology, or just plain risk areas exist in that illustration:

- Oil rings are used to lift oil from the sump into the bearings.

- The back-to-back oriented thrust bearings are not located in a cartridge.

- Bearing housing protector seals are missing from this picture.

- Although the bottom of the housing bore (at the radial bearing) shows the desired passage, the same type of oil return or pressure equalization passage is *not* shown near the 6 o'clock position of the thrust bearing.

- There is uncertainty as to the type or style of constant level lubricator that will be supplied. Unless specified, the pump manufacturer will almost certainly provide the least expensive constant level lubricator

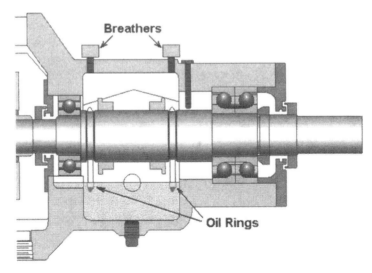

Figure 2.1: A typical bearing housing with several potentially costly vulnerabilities.

configuration. Putting it another way: The best lubricators are rarely found on newly sold pumps.

Each of these issues merits additional explanation and will be discussed in the next chapters. Recall also that our considerations are confined to lubrication issues on process pumps with liquid oil-lubricated rolling element bearings. The great majority of process pumps used worldwide belong to this lubrication and bearing category. Small pumps with grease-lubricated bearings and large pumps with sleeve bearings and circulating pressure-lube systems are not discussed in this text.

ANSI AND ISO VERSUS API PUMPS

ANSI stands for American National Standards Institute, ISO is the International Standards Institute, and API is the American Petroleum Institute. In general, ANSI and ISO pumps comply with dimensional standards; the measurement conventions are inches and millimeters, respectively. ANSI pumps will have the same principal dimensions regardless of manufacturer, as will ISO-compliant pumps in their respective size groups. Principal dimensions for the sake of this overview include, but are certainly not limited to, the distance from base mounting surfaces to the shaft centerline or to the pump suction and discharge flange faces.

The widely used API-610 pump standard is aiming for high strength and reliability. This standard is often used for pumping services that qualify for one or more of the labels hazardous, flammable, toxic, or explosion-prone. The API-610 standard has been called a quality standard, although that should not be viewed as a negative comment on the fitness of ANSI and ISO pumps for safe and long-term satisfactory service. API pumps are centerline-mounted (Figure 2.2); ANSI pumps are usually foot-mounted (as shown earlier in Figures 1.1 and 1.2). The thermal rise of the shaft centerline of a horizontal ANSI pump can be as much as three times greater than that of the centerline-mounted API pump. Thermal rise is taken into account during pump alignment (Chapter 14).

But the API-610 standard should not be viewed as infallible, and the wording in the inside cover of the standard makes that often overlooked point. Reliability-focused users have been justified to deviate from it when experience and technical justification called for such deviations. This text

Figure 2.2: The suction and discharge nozzles on this centerline-mounted API-style pump are upward-oriented. Oil mist lubrication is applied throughout; the coupling guard was removed for maintenance (Source: Ref. 2).

will deal with some issues in which API-610 needs user attention and suitable amendment.

Experience-based selection criteria summarize the proven practices of the Monsanto Chemical Company's Texas City plant in the 1970s (Ref. 2). Among these was the recommendation of using in-between-bearing pump rotors whenever the product of power input and rotational speed (kW times rpm) would exceed 675,000.

For general guidance, Ref. 2 asked that API-610 compliant pumps be given strong consideration whenever one or more of the following six conditions are either reached or exceeded:

- Head exceeds 350 feet (~106 m)

- Temperature exceeds 300°F (~150°C) on pipe up to and including a 6-inch nominal diameter; alternatively, if the temperature exceeds 350 degrees F (~177°C) on pipe starting with an 8-inch nominal diameter

- Pumps with drivers rated in excess of 100 hp (starting at 75 kW and higher)

- Suction pressures greater than 75 psig (516 kPa)

- Flow in excess of the flow at best efficiency point (BEP) for the pump at issue

- Speeds in excess of 3,600 rpm

Exceptions to the six conditions can be made judiciously. To qualify for such an exception, the pumped fluid should be nonflammable, nontoxic, and nonexplosive. In general, exceptions might be granted if the vendor can demonstrate years of successful operation for the proposed pump in a comparable or perhaps even more critical service.

Best-of-class (BOC) pump users are ones that can get long failure-free runs from their pumps. BOCs have on their bidders' lists only vendors (or custom builders) with proven experience records. Such vendors and manufacturers would be well-established and would have a record of sound quality and on-time deliveries.

Exceptions taken by a bidder to the owner-operator's specification would be carefully examined for their potential reliability impact. This examination process serves as a check on the pump manufacturers' understanding of the buyer's long-term reliability requirements. Orders would be placed with competent bidders only.

Because these vendors use a satisfied workforce of experienced specialists, do effective training and mentoring, and have not disbanded their quality control and inspection departments, their products will command reasonable pricing. Reasonable pricing should not be confused with lowest pricing, although reasonable pricing may indeed be lowest in terms of life-cycle costing.

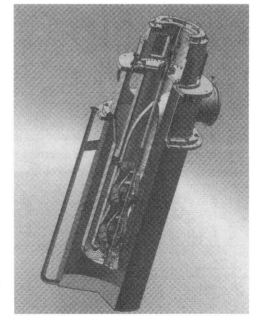

Figure 2.3: Custom-built vertical pipeline pump with drive motor removed (Ref. 3).

Vertical pumps are available in hundreds of styles and configurations, and Figure 2.3 shows a two-stage pump custom-built for pipeline service. This pump is unique because the entire pumping element can be removed as one piece for maintenance. The motor mount (with the motor attached) would be removed first, 24 screws are removed next, and the whole pump lifted out. It is a good example of a design that is user-oriented in terms of maintenance and probable overall reliability (Ref. 3).

Both API and non-API standards are used in custom-built pumps, depending on user preference, type of service, and prevailing experience. Competent designs are available not only from original equipment manufacturers (OEMs) but also from certain key custom design innovators and manufacturing specialists. We count them among the quality providers.

In all instances, the pump owner-operator would compile a specification document that incorporates most, if not all, of the items discussed in this text. The pump owner-operator or its designated project team would mail the document to at least two but more likely three or four of these quality bidders or quality providers. Their replies or cost quotes would be carefully reviewed. These replies would describe the vendor's offer pictorially, and Figure 2.4 is typical of add-ons that the vendor can submit in support of its claim of proven experience and to show pump construction details.

What We Have Learned

*   Getting good pumps requires an up-front effort of defining what the buyer really wants. The user's present or future maintenance philosophy will determine what belongs into a specification.

*   Without a good specification, the buyer is likely to get a "bare minimum" product. Bare minimum products will require considerable maintenance and repair effort in future times.

*   The specification document must be submitted to competent bidders only. Some bidders may ask you to grant a waiver to a particular specification clause; insist they prove that they have understood the reason and purpose of the clause they are unable or unwilling to meet. Never waste your time on bidders that take blanket exception to your entire specification.

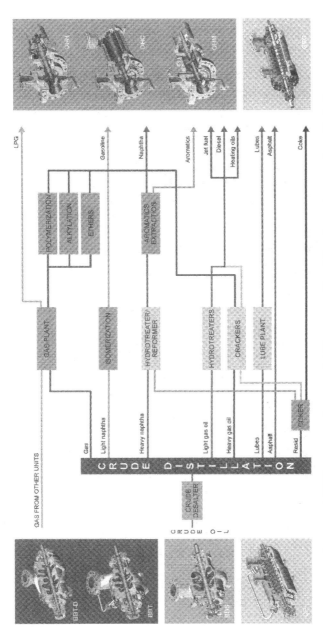

Figure 2.4: HPI process scheme and API pumps offered for the various services (Ref. 4).

- If the owner-purchaser of a process plant grants a waiver to a certain specification clause, then he should understand the extent to which noncompliance will lead to increased maintenance requirements, downtime, or even catastrophic failure risk.

- In the end, you get not what you *ex*pect, but you get what you *in*spect. Inspection is one of the costs of getting reliable process pumps.

References

1. Lubrication Systems Company, Houston, Texas. Photo contributed by and used with the permission of Don Ehlert.
2. Ingram, James H.; "Pump reliability—where do you start," presented at ASME Petroleum Mechanical Engineering Workshop and Conference, Dallas, TX, September 13-15, 1981.
3. Alfred Conhagen Inc., Houston, TX.
4. Sulzer Pumps, Ltd, Winterthur, Switzerland. By permission (2010).

Chapter 3

Foundations and Base Plates

Pumps can be found mounted in many different ways; there are times and places to do it at least cost, and times and places to do it with uncompromisingly high quality.

Plants that use stilt mounting (Figure 3.1) often fall short of achieving the best possible equipment reliability. Best practices plants secure their pumps more solidly on more traditional foundations. Stilt-mounted pump sets lack overall stiffness but have been used for small American National Standards Institute (ANSI) pumps in which the sideways-move capability of the entire installation was thought to equalize piping-induced stresses. Among its few advantages is low initial cost.

However, there are serious shortcomings because stilt-mounting will not allow pump vibration to be transmitted through the base plate to the foundation and down through the subsoil. Proper foundation mounting permits damping of vibration, which can result in a significant increase in mean time between failures (MTBF), longer life of mechanical seals and bearings, and favorably low total life cycle cost (Ref. 1).

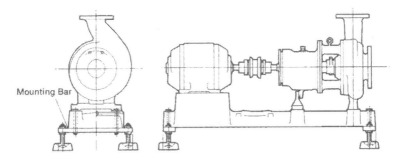

Mounting Bar

Figure 3.1: ANSI pump set on a stilt-mounted base plate (Source: ITT/Goulds, Seneca Falls, NY).

Securing Pumps in Place—With One Exception

Again, proper field installation of pumps has a measurable positive impact on pump life. Even a superb design will give poor results if poorly installed. A moderately good pump design, properly installed, will give strong results (Ref. 2). Proper installation refers to a good foundation design, no pipe strain (see Chapter 4), and good shaft alignment (Chapter 14), to name just a few. No pump manufacturer designs its pumps strong enough to act as a solid anchoring point for incorrectly supported piping or for piping that causes casings and pump nozzles to yield and deflect. Also, pumps have to be secured properly to their respective base plates, and these base plates have to be well-bonded to the underlying foundation. Epoxy grout is used to do this bonding in modern intstallations.

There is one exception, however: Vertical in-line pumps (Figure 3.2) are not to be bolted to the foundation. They are intended to respond to thermal and other growths of the connected piping and must be allowed to float or slide a fraction of an inch in the x- and y-directions. The foundation mass under vertical in-line pumps can be much less than that under the more typical horizontal pump.

Making the foundation mass three to five times the mass of the pump and its driver has been the rule of thumb for horizontal pumps. For vertical in-line pumps, it is acceptable to make the concrete foundation about 1.5 to 2 times the mass of the pump-and-driver combination (Ref. 3).

Figure 3.2: Vertical in-line pumps are not to be bolted to the foundation. They should be allowed to move with the connected pipes.

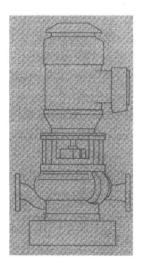

WHY NOT TO INSTALL PUMP SETS IN THE
AS-SHIPPED CONDITION

There are obviously some flaws in the grout surrounding the base plate in Figure 3.3. (However, the equipment owner invested in a modern small oil mist lubrication unit.) Note the hollow space under the electric motor. Lack of support under motors often invites resonant vibration. Rigorous written installation procedures are needed and must be adhered to if long equipment life is to be achieved.

Before delving into other installation matters, note the alignment jacking provisions in Figure 3.4, where the purchaser specified an arrangement that allows insertion (and later removal) of alignment jacking tabs in the x- and y-directions next to each of the four motor feet. (A fixed jacking tab arrangement is shown later—Chapter 14, Figure 14.3.)

Portable jacking tabs, Figure 3.4 (inserted in a welded-on bracket), allow driver alignment moves to be made. Thereafter, the jacking bolts are backed-off and the entire tab is removed. When jack screws are left tightened against the motor feet, motor heat and thermal growth might force the feet into these bolts even more, sometimes causing the entire motor casing to

Figure 3.3: A typical, but obviously flawed, "conventional" pump foundation. (Source: Lubrication Systems Company, Houston, TX).

Figure 3.4: Removable alignment jacking tabs shown inserted in three of four locations next to the two motor feet shown here (Source: Stay-Tru, Houston, TX).

distort (Ref. 1). Note, therefore, that backing-off jacking bolts should be one of many installation checklist items.

To ensure level mounting throughout, the base plate is placed on a foundation into which hold-down bolts or anchor bolts (Figure 3.5) were encased when the reinforced concrete foundation was being poured (Ref. 2). For proper stretch and long life, these anchor bolts (Figure 3.5) must have a diameter-to-length ratio somewhere between 1:10 and 1:12. The anchor bolts are provided with sleeves or other flexible fill. The sleeves prevent entry of grout and accommodate the differing amounts of thermal growth of a concrete foundation relative to that of a steel base plate.

CONVENTIONAL VERSUS PRE-FILLED BASE PLATE INSTALLATIONS

In general, horizontal process pumps and drivers are shipped and received as a "set" or package (i.e., already premounted on a base plate). Seeing a conveniently mounted-for-shipping pump set often leads to the erroneous assumption that the entire package can simply be hoisted up and placed on a suitable foundation. However, that's certainly not best practice, and best-of-class (BOC) plants will not allow it.

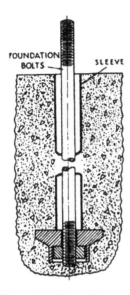

FOUNDATION BOLTS

SLEEVE

Figure 3.5: Foundation anchor bolts and sleeves encased in pump foundation.

The shipping method has little to do with how pumps should best be installed in the field, and pump installation issues merit considerable attention. Again, before installing a conventional base plate, the pump and its driver must be removed from the base plate and set aside. Leveling screws (Figure 3.6) are then used in conjunction with laser-optic tools or a machinist's precision level. With the help of these tools, the base plate mounting pads are brought into flat and parallel condition side-to-side, end-to-end, as well as diagonally, all within an accuracy of 0.001 inches/ feet (~0.08 mm/m) or better. The nuts engaging the anchor bolts are being secured next, and the hollow spaces within the base plate as well as the space between base plate and foundation are being filled with epoxy grout.

The traditional approach to joining the base plate to the foundation has been to build a liquid-tight wooden form around the perimeter of the foundation, and to fill the void between the base plate and the foundation with either a cement-based or an epoxy grout. Both grouting approaches are considered conventional and should not be confused with the preferred epoxy prefilled method which is highlighted subsequently.

Grouting a base plate or skid to a foundation requires careful attention to many details. A successful grout job will provide a mounting surface for the equipment that is flat, level, very rigid, and completely bonded to the foundation system. Many times these attributes are not obtained during the first attempt at grouting, and expensive field correction techniques have to be employed. Predominant installation problems involve voids and distortion of the mounting surfaces. In fact, the most frequently overlooked foundation and pipe support problems are related to foundation settling.

How so? Just as a residential dwelling or sidewalk will probably shift, settle, and crack over time, pump foundations and supports should be expected to do the same. It would be prudent to plan for preventive or corrective action over time or during plant shutdowns. Fortunately, there is now an even better option; it involves the use of epoxy prefilled base plates.

Figure 3.6: Steel base plate with an anchor bolt shown on the left and a leveling screw on the right. A chock (thick steel washer) is shown between leveling screw and foundation (Ref. 2).

EPOXY PREFILLED BASE PLATES

As of about the year 2000, best practices companies (BPCs) have increasingly used "monolithic" (all-in-one, epoxy prefilled) steel base plates in sizes approaching 1.5 m × 2.5 m (about 5 ft × 8 ft). Larger sizes become cumbersome due to heavy weight, see page 29.

In the size range up to about 1.5 m × 2.5 m, conventional grouting procedures, although briefly mentioned in this text, are being phased out in favor of base plates prefilled with an epoxy resin or grout (Ref. 3). These prefilled steel base plates then represent a solid block (the "monolith") that will never twist and never get out of alignment.

The process includes the following successive stages, all done under controlled conditions before shipment to the site:

1. Base plate fabrication (no pour holes are needed for prefilled base plates)
2. Stress relieving
3. Pre-grouting (primer application) in preparation for prefilling. (If there are large pour holes, then the inverted base plate must be placed on a sheet of plywood; Figure 3.7.)

4.   Fill with epoxy grout and allow it to bond and cure
5.   Invert and machine the mounting pads to be flat; and then verify flat-
     ness before shipment (Figure 3.8). Protect and ship (Figure 3.9)—pos-
     sibly even with pump, coupling and driver mounted and final-aligned

The advantage of prefilling is notable. Jobs with pumps in the 750-
kW category and total assemblies weighing more than 12,000 kg (24,600
lbs) have been done without difficulty on many occasions. A convention-
ally grouted base plate requires at least two pours, plus locating and repair-
filling of voids after the grout has cured. Prefilled or pregrouted base plates
travel better and arrive at the site flat and aligned, just as they left the factory. Their structural integrity is better because they do not require grout
holes. Their installed cost is less, and their long-term reliability is greatly
improved.

How to Proceed if There Is No Access to Specialist Firms

If a specialist firm is not available or if upgrading is done at a field
location, then ascertain that the base plate's underside is primed with high-
quality epoxy paint. In general, base plates are specified with an epoxy primer
on the underside. This primed underside should be solvent-washed, lightly
sanded to remove the glossy finish, and solvent-washed again. For inorganic
zinc and other primer systems, the bond strength to the metal should be
determined; expert instructions will be helpful. There are several methods
for determining bond strength, but as a general rule, if the primer can be
scraped off with a putty knife, then the primer should be removed. Sand
blasting to an SP-6 finish is the preferred method for primer removal. After
sand blasting, the surface should first be solvent-washed and then pregrouted
(i.e., epoxy-filled) within 8 hours.

By its very nature, pregrouting a base plate will greatly reduce problems of entrained air creating voids. However, because grout materials are
highly viscous, proper placement of the grout is still important to prevent air
pockets from developing (Ref. 4). The base plate must also be well supported
(Figure 3.7) to prevent severe distortion of the mounting surfaces because of
the weight of the grout.

Once the pregrouted base plate has been fully cured, it is turned right-
side-up, and a complete inspection of the mounting surfaces is performed

Figure 3.7: Underside of a base plate after a prime coat has been applied. It is ready to be filled with epoxy. The large pour holes identify it as an old-style "conventional" base plate being converted to prefilled style (Source: Stay-Tru, Houston, TX).

(Figure 3.8). If surface grinding of the mounting surfaces is necessary, then a postmachining inspection must also be performed. Careful inspection for flatness, coplanarity, and relative levelness (colinear) surfaces should be well documented for the facility's construction or equipment files. The methods and tolerances for inspection should conform to the following:

- *Flatness.* A precision ground parallel bar is placed on each mounting surface. The gap between the precision ground bar and the mounting surface is measured with a feeler gauge. The critical areas for flatness are within a 2- to 3-inch radius of the equipment hold-down bolts. Inside of this area, the measured gap must be less than 0.001 inches. Outside the critical area, the measured gap must be less than 0.002 inches. If the base plate flatness falls outside of these tolerances, then the base plate needs to be surface ground.

- *Coplanarity.* A precision ground parallel bar is used to span across the pump and motor mounting pads in five different positions, three lateral and two diagonal. At each location, the gap between the precision ground bar and the mounting surfaces is measured with a feeler gauge.

Figure 3.8: Flatness and level measurements determine if the now machined pre-filled base plate has been properly machined. It is then ready to be installed on a foundation at site (Source: Stay-Tru, Houston, TX).

If the gap at any location along the ground bar is more than 0.002 inches, then the mounting pads will be deemed non-coplanar, and the base plate will have to be surface ground (Figure 3.8). Outsourcing base plate design, fabrication, and prefilling with epoxy grout has often been found economically attractive. Figure 3.9 shows it ready for shipment.

• *Relative Level (Colinearity)*. It is important to understand the difference between relative level (see preceding bullet points) and absolute level. Absolute level is the relationship of the machined surfaces to the earth. The procedure for absolute leveling is done in the field and is not a part of this inspection. Relative level is an evaluation of the ability to achieve absolute level before the base plate gets to the field.

Conventional grouting methods for nonfilled base plates, by their very nature, are labor and time intensive (Ref. 5). Using a pre-grouted base plate with conventional grouting methods helps to minimize some of the cost, but the last pour still requires a full grout crew, skilled carpentry work, and good logistics. To minimize the costs associated with base plate installations, a new

Figure 3.9: Epoxy pre-filled base plate fully manufactured by a specialty company, shown ready for shipment (Source: Stay-Tru Company, Houston, TX).

field grouting method has been developed for pregrouted base plates. This new method (Ref. 6) uses a low-viscosity, high-strength epoxy grout system that greatly reduces foundation preparation, grout form construction, crew size, and the amount of epoxy grout used for the final pour.

WHAT WE HAVE LEARNED:
CHECKLIST OF FOUNDATION AND BASE PLATE TOPICS

• Use ultra-stiff, epoxy-filled formed steel base plates ("StayTru" method or an approved equivalent) on new projects and on optimizing existing facilities:

(a) Proceed by first inverting and preparing the base plate; use recommended grit blasting and primer paint techniques.

(b) Fill with suitable epoxy grout to become a monolithic block.

(c) Allow to cure; after curing, turn over and machine all mounting pads flat and coplanar within 0.0005 inches per foot (0.04 mm/m).

(d) Next, install complete base plate on pump foundation. Anchor and level it within the same accuracy.

(e) At final installation, place epoxy grout between the top of the foundation and the space beneath the monolithic epoxy prefilled base plate.

- On welded base plates, make sure that the welds are continuous and free of cracks.

- On pump sets with larger than 75-kW drivers, ascertain that base plates are furnished with eight positioning screws per casing (i.e., two screws "jacking bolts" per mounting pad:
  (a) These positioning screws could be located in removable tabs (i.e., tabs slipped into a welded guide bracket) or fixed tabs (i.e., tabs welded onto the base plate).
  (b) Pad heights must be such that at least 1/8-inch (3 mm) stainless steel shims can be placed under driver feet.

- Conventional base plates must be installed and grouted on foundation with pump and driver removed. Only then should pump and driver be reinstalled and leveled.

- Epoxy-prefilled base plates can be installed and grouted on a foundation with pump and driver already aligned and bolted down on the base plate.

References

1. Myers, Richard, (1998); "Repair Grouting to Combat Pump Vibration," *Chemical Engineering Book Series,* Volume 1, 1998; (McGraw-Hill, New York, NY).
2. Barringer, Paul, and Todd Monroe; "How to justify machinery improvements using reliability engineering principles," Proceedings of the Sixteenth International Pump Users Symposium, Turbomachinery Laboratory, Texas A&M University, College Station, TX, 1999.
3. Bloch, Heinz P., and Alan R. Budris; *Pump User's Handbook: Life Extension,* 3rd Edition (2010), The Fairmont Press, Lilburn, GA (ISBN 0-88173-627-9).
4. Bloch, Heinz P., and Fred K. Geitner; *Major Process Equipment Maintenance and Repair,* 2nd Edition (1985), Gulf Publishing Company, Houston, TX. (ISBN 0-88415-663-X)
5. Harrison, Donald M.; *The Grouting Handbook,* (Gulf Publishing Company, ISBN 0-88415-887-X).
6. Monroe, Todd R., and Kermit L. Palmer; "Methods for the Design and Installation of Epoxy Prefilled Base Plates," *Marketing Bulletin,* (1997), Stay-Tru Services, Inc., Houston, TX.

Chapter 4

Piping, Stationary Seals, and Gasketing

PIPE INSTALLATION AND SUPPORT

Unless piping configurations and associated layout are done by thoughtful design, things often go wrong. A distinction must be made between the implementation tasks assigned to pipe fitters and the reliability-focused engineering tasks assigned to piping designers.

Accurate computer models and suitable software are used by reliability-focused plants. These plants then obtain the lowest stress, most cost-effective, and least-risk pipe installation by design, and not by any of the less dependable means.

Based on computer recommendations, fixed and sliding pipe supports are located so as not to let thermal movements cause undue loads on pump nozzles. At some locations, the pipe must be suspended by tension springs; at other locations, the pipe must rest on suitably restrained compression springs. The weight of insulation and sometimes even ice or snow loads must be taken into account by the designer.

SLIDING SUPPORTS AND INSTALLATION SEQUENCE DESERVE SPECIAL ATTENTION

There is a widely overlooked item on sliding supports; Steel-Teflon-Steel is not usually a satisfactory sliding support. The steel plates ("shoes") will oxidize and, because of surface roughness, will dig into the Teflon. The best practice is to use Steel-Teflon-Teflon-Steel; in which case, Teflon will slide on Teflon with considerable ease, providing long-term satisfactory sliding action.

Best practice requires installation of piping from vessels or other structures upstream of the pump and towards the pump suction nozzle, to a point about 10-15 feet (3-5 m) from the suction nozzle. Then, one places a pipe flange at the suction nozzle and works toward the pipe run that had been terminated 10-15 feet upstream.

A similar work sequence is next used on the downstream piping. On the downstream pipe installation, one would install pipe from receivers, destination vessels or other structures downstream of the pump and install pipe toward the pump discharge nozzle. This downstream pipe installation sequence should initially terminate at a point about 10-15 feet (3-5 m) from the pump discharge nozzle. Then, one places a pipe flange at the discharge nozzle and works toward the pipe run that had been terminated 10-15 feet downstream.

Upstream and downstream of the process pump, the final connections are then made either at gasketed pipe flanges or by welding. While making these final connections at both upstream and downstream terminations 10-15 feet from the pump, dial indicators set up to monitor pump nozzle and bearing housing movement must not show displacements in excess of 0.002 inches (0.05 mm).

MONITORING PIPE STRESS WHILE BOLTING UP

Pumps are designed to allow only limited loading of pump suction and discharge nozzles. Misaligned pipes can produce forces and moments on pump nozzels that vastly exceed maximum allowable values. Excessive piping loads can cause high vibration, shaft misalignment, seal distress, bearing overload, and coupling failures.

To keep within the allowable limits, several dial indicators are set up to monitor the pump's sensitivity to pipe stress. Dial indicator movement is monitored while initial and final bolt tightening is in progress. Four dial indicator stems are set to contact pump and driver feet to detect unsupported (soft-foot) conditions; two additional dial indicators observe the pump bearing housing for movement in the x and y-directions. Any indicator needle displacement in excess of 0.002 inches (0.05 mm) will require corrections to the piping.

Again, a process pump should never be allowed to serve as a pipe support. Chain-falls or come-along hoists and other supplementary mechanical tools (pulling devices) are never allowed or used by reliability-focused pump installation crews.

In general, maximum misalignment pipe flange to pump nozzle and flange-to-flange should be kept within the limits of Figure 4.1. Before allowing connections to be made, the two mating faces should be

1. Parallel with each other within 1/32 inches (0.8 mm) at the extremity of the raised face (i.e., "A" and "B" in Figure 4.1 should differ by no more than 1/32 inch (0.8 mm).

2. Concentric so their centerlines coincide within 1/8 inch (i.e., the offset "[C minus D]" should not exceed 1/8 inch (3 mm).

There are two important, simple, yet useful tests:

1. It must be possible for an average-size worker to push the misaligned piping into place with his two hands, without using any supplementary "come-alongs" or other mechanical tools.

2. Once the gasket has been inserted and the bolts have been torqued up, dial indicators observing the upward and sideways motion at the pump suction and discharge nozzles cannot move in excess of 0.002 inches (0.05 mm) in any direction.

FLANGE LEAKAGE

Although not always considered the pump person's responsibility, flange leakage issues must be understood by the pump person.

Table 4.1 lists nine causes of flange leakage that are most often considered. These nine causes can be separated into the following categories: design-related and installation or system-related.

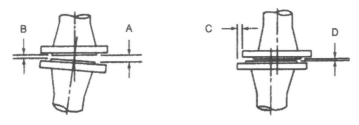

Figure 4.1: Limits of flange deviations (Ref. 1).

Table 4.1
Nine prominent causes of flange leakage

Uneven bolt stress
Misaligned or tilted flange
Nonconcentric installation of gasket
Dirty or damaged flange faces
Excessive bending moments imposed by the connected piping
Thermal shock during operation
Incorrect gasket size or material
Incorrect flange facing
High vibration levels

Most flange leakage problems have to do with issues brought on by flawed work processes in the field. Problems are rarely the result of basic gasket deficiencies or selection mistakes.

WHAT TO DO PRIOR TO GASKET INSERTION

Rules of thumb in regard to flange condition. The following rules are rather subjective and really just follow plain common sense:

- Check condition of flange faces for scratches, dirt, scale, and protrusions. Wire brush clean as necessary. Deep scratches or dents are those that are deeper than 0.005 inches (0.12 mm) or extend over more than 1/3 of the width that will later be taken up by the gasket. These will require refacing with a flange facing machine.

- Check that flange facing gasket dimension, gasket material and type, and bolting are per specification. Reject nonspecification situations. Note also that picking an improper gasket size is a common error.

- Check gasket condition; only new gaskets should be used. Damaged gaskets (including loose spiral windings) should be rejected. The inner diameter (ID) windings on spiral-wound gaskets should have at least three evenly spaced spot welds or approximately one spot weld every 6 inches of circumference (see American Petroleum Institute [API] 601).

- Use a straightedge and check facing flatness. Reject warped flanges.

- Check alignment of mating flanges. Avoid use of force to achieve alignment—remember the two-hand rule given previously.

Joints not meeting these criteria should be rejected, and defect type, cause, and remedial action should be mapped out (see Tables 4.2 and 4.3).

Spiral Wound and Kammprofile Gaskets

Spiral wound gaskets were developed to improve performance in high-pressure applications ranging from flanged pipe connections to heat exchangers. They consist of alternating plies of compressible filler material and a thin-gauge metallic strip wrapped like the grooves in a phonograph record. Spiral wound gaskets provide the needed pressure resistance in these applications (Figure 4.2). In the 1980s, alternative materials such as flexible graphite replaced asbestos as the filler in these gaskets; yet their basic design has remained unchanged since they were invented in the early 1900s.

Today the most common method for centering a spiral wound gasket makes use of a metal outer ring. The outer guide ring serves to center the gasket in the flange and to limit its compression. If the sealing surfaces are compressed against this centering ring (and no inner ring is present), then a metal-to-metal seal may be formed. This is acceptable, provided the flanges remain at a steady temperature. However, when gasket assembly stress cannot be adjusted to accommodate upset conditions or thermal cycling, the seal may be subject to premature failure. This is especially true when graphite fillers are used without inner rings. In addition to its performance-related functions, the outer guide ring also serves to identify the size, pressure class, and material composition of the gasket.

Both spiral wound and the relatively new kammprofile (derived from the German "Kamm" = comb) gaskets are used extensively in refineries and petrochemical plants. They are primarily serving in applications subject to thermal cycling, pressure variations, flange rotation, stress relaxation, and creep. As of the late 1990s, there has been a discernable shift away from the use of spiral wound gaskets in favor of kammprofiles. Kammprofile gaskets (Figure 4.3) tend to provide better sealing performance and longer service life but will cost more (Ref. 2).

Table 4.2: Defect Causes and Remedies

Type of Defect	Defect Cause	Possible Remedial Steps
a) Excessive gasket extrusion	Excessive seating stress	a) Select replacement material with better cold flow properties. b) Select replacement materials with su=perior load-carrying capacity
b) Gasket excessively compressed	Excessive gasket stress	a) Select material with better load-carrying capacity/higher seating stress b) Select thinner gasket c) Increase contact area of gasket d) Reduce number of bolts e) Redesign flange, if necessary
c) No gasket compression achieved	Insufficient applied gasket pressure or bolting stress	a) Select gasket material requiring lower seating stress b) Select thicker gasket cross section c) Reduce gasket area to allow higher unit seating load d) Apply additional torque e) Bolts should be tightened in sequence f) Gasket may be relaxed because of operating temperature. Tighten again after pipe reaches operating temperature g) Ensure threads are sufficiently long; nuts must make face contact h) Increase number of bolts, if possible i) Increase diameter of bolts Change to high tensile bolt materials (not recommended in H_2S-containing environment)
d) Gasket is badly corroded	Gasket material is incompatible with process fluid	a) Select or upgrade to replacement material with improved corrosion resistance
e) Gasket is mechanically damaged because overhang of raised face or flange bore	Wrong placement of gasket	a) Make sure that gaskets are properly centered in joints b) Review gasket dimension to ensure gasket is properly sized
f) Gasket is thinner on OD than on ID	Flange deficiency or excessive bending	a) Select gasket dimension as to move gasket reaction force closer to bolts; bending would be minimized b) Select a softer gasket material to lower required seating stress c) Reduce the gasket area as to lower seating stresses
g) Gasket unevenly compressed in circumference	Uneven load on gasket	a) Ascertain proper sequential bolt-up procedures ("criss-cross" sequence) are being followed

Table 4.3: Diagnosing gasket defects (Ref. 2)

Undercompression

Cause	Effect	Solution
Insufficient torque	Filler not conformed to sealing surfaces Premature leakage	Increase torque to increase gasket stress or reduce winding crosssectional area
Insufficient available bolt force	Filler not conformed to sealing surfaces Premature leakage	Reduce crosssectional area or use a kammprofile
Filler density too high	Problems sealing at low stud loads Leaks can develop if windings take the initial load and the graphite is underloaded	Address gasket design with manufacturer

Overcompression

Cause	Effect	Solution
Excessive torque/available bolt force	Radial buckling (especially with gaskets with no inner rings) of the windings and/or inner ring Process stream contamination/leakage	Reduce torque (see gasket manufacturer)
Low density winding—flanges contact outer guide ring	Reduced stress within the windings Leakage because gasket cannot be loaded properly Gasket will seal if compressed sufficiently	Address gasket design with manufacturer
Filler density too high	Outer guide ring cups, warps, or tilts Can cause inner ring to buckle and excessive guide ring roll	Address gasket design with manufacturer

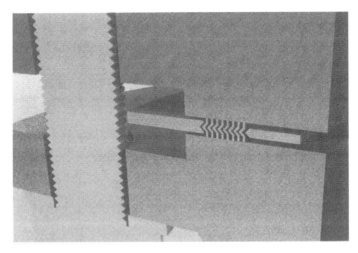

Figure 4-2: Spiral-wound gaskets are reinforced with metal rings to prevent buckling in service and damage from improper handling.

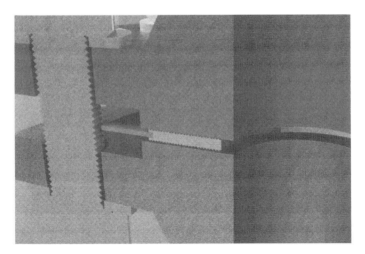

Figure 4.3: Kammprofile gaskets consist of a solid metal core with concentric serrations and faced with a nonmetallic material such as flexible graphite or various grades of polytetrafluoro-ethylene (PTFE).

Kammprofile pipe flange gaskets compress significantly less than spiral wound gaskets—on the order of 0.022 inches compared with 0.030 inches to 0.075 inches for a spiral-wound gasket. This means kammprofile gaskets load more quickly with less risk of nonparallel flanges. One disadvantage is that the graphite facing is more susceptible to mechanical damage if not properly handled. Because the graphite is not protected by the windings as it is in spiral wound gaskets, it also can be damaged by oxidation at temperature between 600°F and 800°F (315-427°C) depending on the grade of graphite. (Higher temperatures may be possible by including a mica-based layer around the OD to protect the graphite.) It is best to specify good quality, inhibited graphite when using these types of gaskets.

PIPE, HYDRAULIC TUBING, OR FLEXIBLE CONNECTIONS?

Considerable precautionary care is needed when contemplating the use of anything other than pipe. In theory, it would be cost-effective and quick to install hydraulic tubing for auxiliary fluid systems such as pump seal flush and lubrication supply lines. However, even stainless steel tubing rated for high pressures is not used in process pumps. Instead, "hard pipe" is employed throughout. Hard pipe is less likely to bend or deflect when inadvertent contact is made with tools, equipment, or personnel. As to what is adequate, the best gage is a worker's eye: There should be no visible movement when his full body weight is applied to auxiliary piping.

A somewhat analogous situation exists with flexible hose and expansion joints. These, too, are not recommended for process pumps. Flexible joints might potentially alleviate pipe strain but would certainly require expert and careful installation.

Flexible and expansion joints are never used in flammable, toxic or explosive services by reliability and safety-focused personnel. If there should be a fire at or near a pump equipped with flexible hose or expansion joints, then these weak links will likely be the first to let go, and the risk of aggravated failure and disaster would increase exponentially.

Sticking with pipe on process pumps is prudent. The pipe must be properly installed, and no dangerous short cuts should ever be allowed. Freestanding or small pipe must always be braced or gusseted with a diagonal bracket or other suitable two-plane bracing or support (Figure 4.4). Equipment vibration tends to weaken unsupported free-standing pipes.

Never use pipe for handrails. In the mid-to-late 1800s, pipe was widely used for handrails on elevated structures in the mining industry. Pipe had become readily available and the human hand can grasp handrails rather comfortably. To prevent rusting, the pipe was painted. But rust formed on the inside of the pipe and, unbeknownst to some workers, the pipe walls got progressively thinner. More than once, a worker leaned against the nicely painted but now weakened handrail and fell to his death. That is why pipe is not used for handrails in modern industry.

GUSSETING

Bracing and gusseting are important means of ensuring valve and instrument connections of proper strength and with the lowest possible risk of vibration-induced failure. Two-plane gusseting near a pipe is shown in Figure 4.4; installations at the pump piping must be sufficiently far away to satisfy hand clearance and maintenance access requirements. Single-plane gusseting is not sufficiently vibration resistant; it should not be used.

At all times, the requisite material, welding technology and postweld heat treatment (PWHT) requirements must be observed as well. In general, the involvement of a competent metallurgist should be sought in the design of seemingly insignificant auxiliary piping for process pumps. Suffice it to say that careless piping and support routines are disproportionately responsible for many pump failure incidents.

CONCENTRIC VERSUS ECCENTRIC REDUCERS

Piping reducers are generally installed at the process pump suction nozzle to transition from the larger diameter (low flow velocity, moderate friction loss) suction pipe to the pump suction

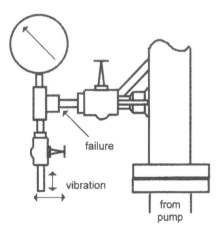

Figure 4.4: Typical small bracing (gusseting) applied in two planes to a small valve (Ref. 3).

nozzle. They should be installed in such a manner that trapped air or vaporized product will not accumulate in any portion of the pipe reducer.

Figures 4.5 and 4.6 serve as installation guidelines of interest.

VIBRATION PROBLEMS IN PIPING

Vibration problems in piping can occur on new installations, or on existing systems with abnormal operating conditions. Surging, two-phase flow, unbalanced rotors, fluid pulsations, rapid valve closures, local resonance, or acoustic problems can cause excessive vibration. A new processing plant can have many miles of piping, and the designers use guidelines

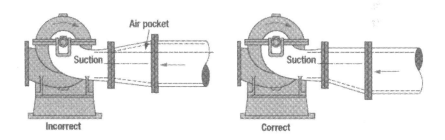

Figure 4.5: In long, horizontal suction pipe runs, air pockets are avoided by using the eccentric reducer (right side of image) with the flat side up.

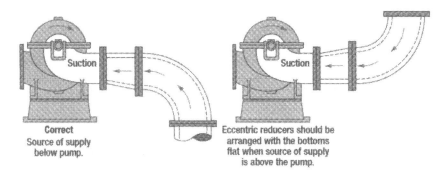

Figure 4.6: If inlet flow originates above the pump suction nozzle, then air or vapor pockets are avoided by using the eccentric reducer (right side of image) with the flat side down.

and experience to know where and how to support the piping. In most new systems, it is necessary to "walk the line" to see what's shaking. Vibrating lines are located, marked, and modified by adding gussets, supports, hangers, or hydraulic shock absorbers (also called "snubbers").

Figure 4.7, "historical piping vibration failures," is based on in-service experience with failed piping (Refs. 3 and 4). This graph is not a design guide because the database is small. Nevertheless, Figure 4.7 represents 22 failures with several similar incidents shown as single points. It is tempting but inadvisable to draw acceptable versus unacceptable limits on this graph. Assuming no failures will occur in regions where no failures are shown would be a dangerous assumption with this limited amount of data.

However, locating a point vibrating at 500 mil p-p and 1 cps with conditions similar to point 7 should certainly make one nervous, and it would not be reasonable to expect a long life from this connection. Point 6 was a 2-inch pipe welded to a ¼-inch-thick shell with the pipe vibrating. This caused an "oil canning" effect (actually a perceived waviness) on the thin metal, and it failed in fatigue in the base metal, not in the weld.

It is reasonable to say that all of these failures could have been avoided by supporting the piping correctly. Good support or gusseting in a vibrating system is mandatory. At a frequency of 15 cps, more than 1 million fatigue cycles can develop in a single day. Without gusseting, the question will not be if but when will fatigue failure occur.

What We Have Learned on Piping and Gasketing Topics

- Do not allow piping to be pulled into place by anything stronger than a pair of human hands. Pipe flange connections requiring chain falls or other mechanical pulling devices are flawed and not allowed by reliability-focused users.

- While making pipe flange to pump nozzle connections, ascertain that no dial indicator moves more than 0.002 inches (0.05 mm) while tightening or loosening flange bolts.

- Observe "piping away from pump" rules.

- Disallow expansion joints and other weak elements in flammable, toxic, or explosive services. In case of fire, the results can be truly catastrophic.

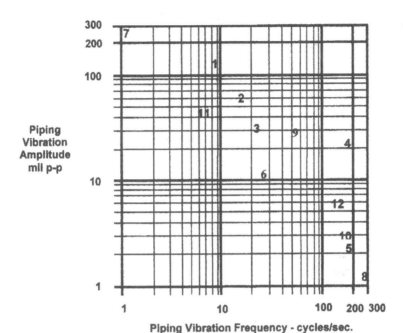

Piping Vibration Frequency - cycles/sec.

Details on Failure Points:
1. 24 inch line connected to vessel, 4 ft. from weld at vessel
2. 8 inch unsupported line connected to 36 inch header, 3 ft from header
3. 12 inch compressor piping failure at support weld
4. 2 inch socket weld 3 ft from socket, no gusseting
5. Crack in weld due to "buzzing" of 2 inch screw compressor line, 2 ft from weld
6. 2 inch pipe to ¼ inch thin wall structure, failed at weld to wall, due to wall flexing
7. Reinforced 5 inch branch connection to 8 inch, 4 ft from weld
8. ¾ inch socket weld, crack at root, lack of fusion, 3 ft. from socket weld
9. 8 inch compressor discharge nozzle 2 ft from weld
10. 6 inch branch connections, flange leaks, weld cracks
11. 1 inch socket weld crack, measured 2 ft from weld
12. ½ inch thread / nipple to gauge failure, in thread, 8 inch from gusset

Figure 4.7: Historical vibration-related piping failures (Refs. 3 and 4).

- Use Steel-Teflon-Teflon-Steel for dependable sliding. Do not allow only a single layer of Teflon (i.e., Steel-Teflon-Steel).

- Ascertain acceptable flange condition and do not allow any compromises on quality of workmanship or on accepted sealing element dimensions.

- Spiral wound and kammprofile are the most widely used gaskets types. Either type is centered by its metal outer ring contacting the flange bolts.

- Small-bore piping must be braced or gusseted in two planes.

- Eccentric pipe reducers must be installed with the flat portion properly oriented. In some instances, the flat side is down; in other cases it should face down.

References

1. Bloch, Heinz P. and Fred K. Geitner; *Machinery Component Maintenance and Repair*, 3rd Edition (2004), Gulf Publishing Company, Houston, TX. (ISBN 0-7506-7726-0).
2. Yoder, Chad, and David Reeves; "Spiral wound or kammprofile gaskets?," *Hydrocarbon Processing*, September (2010).
3. Sofronas, Anthony; *Analytical Troubleshooting of Process Machinery and Pressure Vessels* (2006), John Wiley & Sons, Hoboken, NJ, p.139. (ISBN-13:978-0-471-73211-2).
4. Sofronas, Anthony; "Piping vibration failures," *Hydrocarbon Processing* (August 2002).

Chapter 5

Rolling Element Bearings

A few of the hundreds of bearing styles and sizes successfully used in process pumps are shown in Figure 5.1. Each style and configuration incorporates important features and details that should not be overlooked. Of course, misapplications and misunderstandings relating to bearing technology can lead to costly repeat failures. Paying attention to detail yields the lowest cost of ownership over the long run, and being detail oriented is an important attribute of best-of-class (BOC) performers.

Rolling element bearings are precision components that must be treated with great care. Thoughtful selection and picking the right bearing for the application will allow rolling element bearings to run flawlessly for 6 or more years in centrifugal process pumps.

The material in this chapter summarizes portions of Refs.1 through 6. It emphasizes bearing life extension strategies consistently pursued by suc-

Figure 5.1: Some of the many styles and sizes of rolling element bearings found in process pumps (Refs. 1 and 6).

cessful best-of-class pump owners. This chapter also describes certain practices that deprive many pump owners of longer term reliable operation of
pump bearings.

Bearing Selection Overview and
Windage as a Design Problem

Five typical bearing styles or categories are shown in Figure 5.2. The
pros and cons of each style are often indicated in the respective alphanumeric designations, the certain suffixes to the identification code, or even the
names given to certain bearings. The various designators are etched into the
wide face of the bearing outer ring, and because subtle differences in these
designators are often important, one cannot afford to disregard them.

The cylindrical roller bearing of Figure 5(a) is used in bearings with
high radial loads. However, many different versions of cylindrical roller bearings exist and the alphanumeric suffixes on the designation code of such
bearings are important.

Spherical roller bearings of the type shown in Figure 5.2(d) incorporate
an oil application passage in the center of the outer ring. The lube oil supply
is divided, and equal portions of oil are guided to each row of rolling elements.

Because of the inclined cage in the angular contact bearing of Figure
5.2(e), a certain pumping action exists from the "a" to the "b" direction.
In some instances, lubricant finds it more difficult to flow from "b" to "a"
because windage will have to be overcome. Windage is the fan effect that
generates air flow from the smaller to the larger cage diameter. In essence, an
inclined cage acts as a tiny fan or blower.

In an inadequately designed bearing housing, this windage can cause
fluctuating oil levels or leakage from housing seals. Issues of windage and
pressure balance are related to cage location and configuration; both are important and will be revisited later. (The location of a bearing cage relative to
other bearing components is shown on the bearing nomenclature drawing,
Figure 5.4.)

Generally, the "radial" bearing in a centrifugal process pump is on the
side away from the shaft end where the coupling will be attached (Figure
5.3). With few exceptions, this would place the radial bearing on the housing
side closest to the mechanical shaft seal or pump impeller. In contrast, the
bearing or bearing set commonly called "thrust bearing" is usually located on

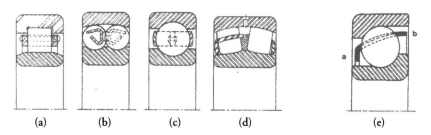

(a) (b) (c) (d) (e)

Figure 5.2: Cylindrical roller bearing (a); self-aligning ball bearing (b); deep-groove ball bearing (c); spherical roller bearing (d); angular contact ball bearing (e)

the side adjacent to the shaft coupling.

Figure 5.3 depicts an American Petroleum Institute (API)-style process pump. API-style pumps are of sturdy construction and are recommended by the American Petroleum Institute (in its Standard API-610) for flammable, toxic, or otherwise hazardous services (Ref. 2). It should be noted that, in API-style process pumps, two separate angular contact ball bearings (see Figure 5.2(e)) make up the set typically used to absorb the axial thrust load created by impeller hydraulics. These bearing sets will be discussed later in this chapter.

Visualize the oil level in a bearing housing reaching just over the side face at a bearing outer ring near the 6-o'clock position (see Figures 5.4 and 5.5). Now visualize the oil level going down by a small distance because of windage, or because of nonequal pressures on the two bearing faces. Suddenly, no more oil will reach the rolling elements and the top layer of oil will overheat and turn black.

Black oil, then, can sometimes be traced to a particular bearing cage geometry or omission of an equalization passage. Black oil in pump bearing housings is either overheated oil or lubricant contaminated by O-ring debris (see also Chapter 8). The reason for oil discoloration must be identified before the proper remedial action can be planned. Opting for just an oil change will not address the root cause of lubricant degradation or discoloration. It will likely result in a repeat failure.

Radial Versus Axial (Thrust) Bearings

In pumps designed and marketed in the United States, the radial bearing is usually configured as illustrated in Figure 5.4. However, European pump designs generally favor the higher load-rated cylindrical roller bearing

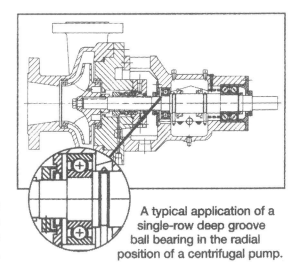

Figure 5.3: API 610 style centrifugal pump cross section with the radial bearing circled.

A typical application of a single-row deep groove ball bearing in the radial position of a centrifugal pump.

of Figure 5.2(a). A higher initial cost and the need for more careful assembly are the distinguishing characteristics of cylindrical roller bearings as compared with typical ball bearings. Regardless of bearing style, the bearing in the radial location should be free to move axially, whereas the outer rings of a thrust bearing assembly should be restrained in place. However, applying an excessive clamping force would risk distorting or buckling the outer ring. Allowing the thrust bearing set to move axially as much as 0.002 inches (0.05 mm) ensures there is no unduly large clamping force.

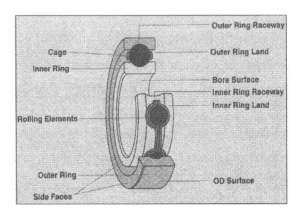

Figure 5.4: Nomenclature for a typical radial ball bearing.

Although the process is called thrust bearing, bearings at the thrust location in pumps are generally absorbing loads in both the axial and the radial directions. A double-row thrust bearing is shown in the American National Standards Institute (ANSI, standardized dimension) process pump of Figure 5.5. The thrust bearing in this illustration is a double-row angular contact bearing (DRACB) with a single inner ring. A double-row bearing with two separate inner ring halves is available for applications in which somewhat higher loads must be accommodated (Figure 5.6). The castellated (cog-type) clamping nut and tab washer in Figure 5.5 (also shown later in Figure 5.18) are required to secure the two inner rings of Figure 5.6 to the shaft.

The castellated clamping nut must be tightened with a special spanner wrench; regrettably, this wrench is not usually found in the average machine shop. Some mechanics tends to use a chisel, which inevitably causes damage to the equidistant cogs ("castellations") in the periphery of the nut. Anyway, a proper spanner wrench must be used and the tab washer discarded after each disassembly. Re-bending a tab would weaken it to the point of risking low-cycle fatigue failure.

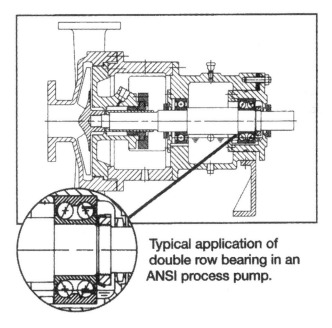

Typical application of double row bearing in an ANSI process pump.

Figure 5.5: ANSI style centrifugal pump cross section with double-row thrust bearing circled. Note castellated nut and tab washer.

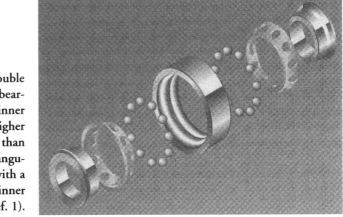

Figure 5.6: Double row angular bearing with two inner rings have a higher load capability than double-row angular bearing with a single, wide inner ring (Ref. 1).

In the double-row two-piece inner ring bearing of Figure 5.6, each inner half incorporates its own land and raceway, but the overall external dimensions are identical to those of a DRACB with a single-piece inner ring. The inner rings of these thrust bearings are clamped to the shaft, and the outer ring is restrained in its housing bore position. Again, the clamping force should be light as to not distort the bearing outer ring. Alternatively and to simply ascertain that the clamping force is not excessive, one might allow an outer ring axial movement of up to 0.002 inches (0.05 mm) relative to the bearing housing bore.

OIL LEVELS, MULTIPLE BEARINGS, AND DIFFERENT BEARING ORIENTATIONS

Attention should be given to oil levels maintained in pump bearing housings; note how the levels differ in Figures 5.3 and 5.5. In Figure 5.3, the oil level is well below the lowermost point of the bearings. However, in Figure 5.5, the oil level is adjusted to reach the center of the lowermost bearing ball. Selecting the right oil level is important (Ref. 3); what level to choose will be discussed later in this chapter.

The thrust bearing set of the API pump in Figure 5.3 is enlarged in the back-to-back layout of Figure 5.7(b). Thrust bearings in pumps are usually back-to-back oriented. The "back" of an angular contact bearing is the wider outer ring land; the narrower outer ring land is the "face." Therefore, Figure

5.7(c) is a "face-to-face" mounted thrust bearing set. Care must be taken to not allow interference at the radii r_a and r_b.

API-610 asks for the contact angles in each bearing making up a set to be equal, which is why we showed them as equal in Figure 5.7. However, two angular contact thrust bearings with equal load carrying capacities are not necessarily best for pumps that clearly experience thrust reversal at start-up only. In such pumps, the then normally unloaded bearing may skid, whereas only the loaded bearing is rolling. Skidding bearing elements (Figure 5.8) wipe off the oil film and can create destructively high temperatures as metal now contacts metal. True rolling, of course, is the design intent for all rolling elements.

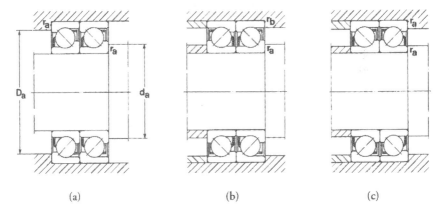

(a)   (b)   (c)

Figure 5.7: Sets of thrust bearings with different orientations (Ref. 4): tandem, for load sharing of a pump shaft thrusting from right-to-left (a); back-to-back, the customary orientation with thrust load on pump shafts expected in each direction (b); face-to-face, rarely desirable in centrifugal process pumps (c).

Again, using only 40-degree sets of angular contact thrust bearings may not optimize bearing life. A close review of API-610 guidelines will show them to be meant as minimum (general) guidelines and not mandatory requirements. In some cases, API-610 represents only the prevailing consensus, or a commendable effort at standardization. Although some of these efforts are unquestionably beneficial, others will be much less so. The foreword and special notes in the API standards encourage users to procure more reliable components or configurations whenever these are available. Our point: Better bearings are available.

The desirable performance characteristics of bearing sets may vary for different styles of pumps. Figure 5.9 shows but a small portion of the many different options and possibilities. For instance, sets consisting of two 15-degree back-to-back angular contact bearings are often best for hydraulically balanced and lightly loaded pumps operating at high speeds. Pumps involving heavy primary thrust loads sometimes use a triplex set; two 40-degree bearings are installed in tandem, and these are then mated, back-to-back, with one 15-degree bearing (Figure 5.10).

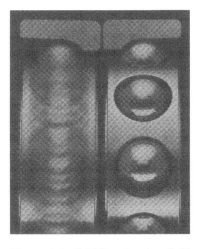

Figure 5.8: Skidding bearing (left) versus rolling bearing (right). Skidding generates heat and quickly damages a bearing

Whenever two or three bearings are mounted adjacent to each other, lubricant application concerns will take on greater importance. Certainly, a small amount of oil applied near the edge of the first bearing might not easily travel to the edge of the third bearing. Likewise, application of a drop of oil at the edge of the third bearing might not readily induce this lubricant to flow toward the first bearing.

Remember "windage" and recall the need to have pressure equalization at all locations inside a bearing housing. "All locations" means spaces to the right and to the left of the radial bearing, as well as spaces to the right and to the left of the thrust bearing.

Matched sets (Figure 5.10) are precision-ground to have the exact (matched) dimensions needed for maximum life and optimum performance. An appropriate match may not necessarily mean identical load angles. Instead, it implies optimizing velocities at the points in which rolling elements make contact with raceway contours. Dimensions and geometries could be nonsymmetrical because the bearings are designed for skid avoidance (Figure 5.11). Matched sets are labeled and sold by knowledgeable bearing manufacturers; the sets come in boxes. Buying these bearings as loose items, from different manufacturers, or from the lowest bidder will increase failure risk and result in a higher ultimate cost of ownership. Remember that process pump bearings are not the same as bearings for cheap roller skates.

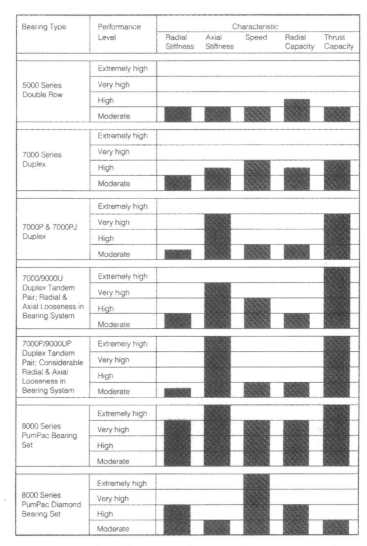

Figure 5.9: Relative bearing performance characteristics (Ref. 5)

Much pertinent advice can always be provided by the application engineering groups of competent bearing manufacturers. Pump users have access to these application engineers by making it a practice to procure bearings from only the most qualified vendors or manufacturers. The premium paid for such bearings is easily justified; it will quickly show up as failure avoidance and pump life extension. The payback will be huge.

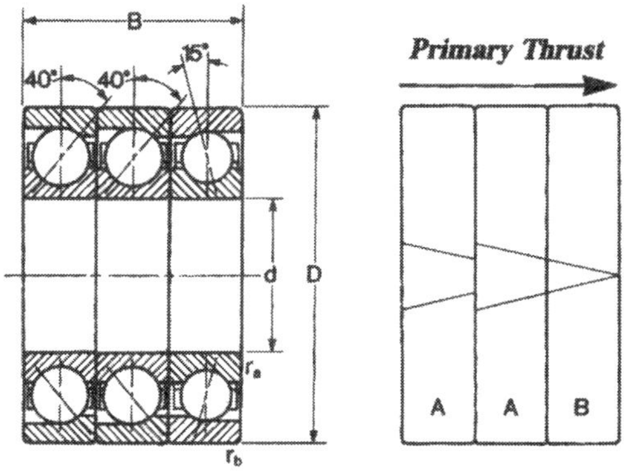

Figure 5.10: Triplex bearing set, consisting of dimensionally matched angular contact bearings. Direction of primary thrust is inscribed.

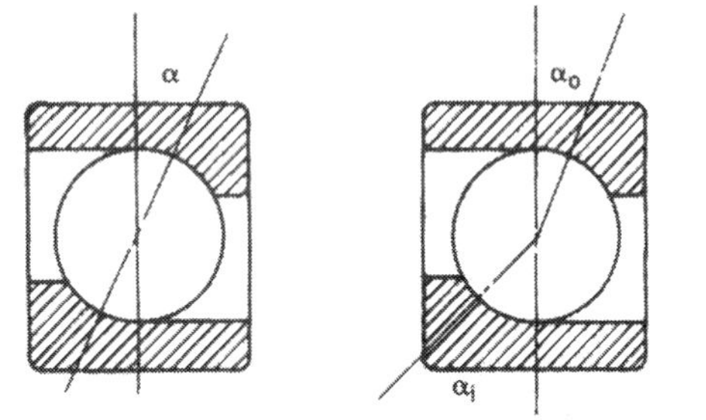

Similar contact angles, inner and outer ring.

Variation in contact angles, inner and outer ring.

Figure 5.11: Competent bearing manufacturers design contact angles to ascertain favorable rolling motion and minimize skidding (Ref. 1).

Upgrading and Retrofit Opportunities

Bearing manufacturers with research and development capabilities frequently and timely respond to users' needs by developing retrofit kits. These needs tend to incorporate improved bearing geometries or advanced materials technologies that include ceramic ("hybrid") bearings. The 40-degree/15-degree matched set of Figure 5.12 was developed for that reason; it most certainly solved a problem that the pump manufacturer and user could not solve without the bearing manufacturer's help.

Another of many retrofit developments is highlighted in the two slightly different double-row angular contact bearings of Figure 5.13. In the event of high shaft loading from right to left, the mounting method depicted in the top sketch will prove more secure than that using a bearing snap ring in the lower portion of Figure 5.13 (Ref. 6). Experience shows the remaining land at the bearing snap ring groove to be narrow and, potentially, too weak

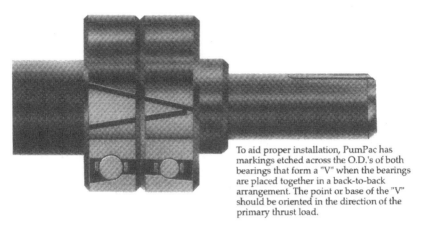

To aid proper installation, PumPac has markings etched across the O.D.'s of both bearings that form a "V" when the bearings are placed together in a back-to-back arrangement. The point or base of the "V" should be oriented in the direction of the primary thrust load.

Figure 5.12: This MRC "PumPac" thrust bearing set with unequal load angles and ball diameters was developed to avoid skidding of the normally unloaded side while maximizing axial load capability in the primary direction of thrust. Machined brass cages are used.

to restrain high imposed axial shaft loads.

Double-row angular contact bearings with two separate inner rings were mentioned earlier in this chapter; note, however, that their use would require applying a clamping torque (threaded shaft end and provision of a castellated clamping nut) to keep the bearing properly assembled.

BEARING CAGES

Bearing cages are needed to keep rolling elements equidistant from each other. The four configurations illustrated in Figures 5.14 and 5.15 are but a small sample of the many geometries and materials available for ball bearing cages. Although discouraged by API-610, steel cages are occasionally found in process pumps. Steel cages are generally less forgiving than machined brass cages. In fact, whenever rivets are used to keep together the mirror-image halves of certain pressed steel bearing cages (see Figure 5.4), the rivet heads are considered a weak link. Should they pop off, massive failure will often result.

Machined brass cages (see also Figures 5.8 and 5.15) are considered more tolerant of minor installation defects and lubrication deficiencies. But

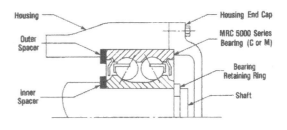

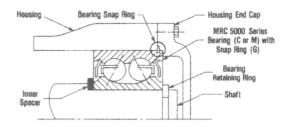

Figure 5.13: Double-row angular contact bearing retained and located with spacer and step in bearing housing (top) versus snap ring secured in bearing outer ring groove and held in place by housing end cap (bottom). The inner rings of these bearings will have to be mounted with a shaft interference fit of approximately 0.0005-0.0007 inches (12-17 µ).

they are not immune to damage should such deficiencies develop. Machined brass cages may respond poorly to situations in which skidding is involved.

For years, it had been argued that permitting polyamide cages in process pumps would be out of tune with the safety and reliability improvement goals professed by many users. But modern polyamide cages include Nylon 46 and 66, Kevlar and linear-poly-phenylene sulfide (L-PPS). L-PPS has excellent thermal and chemical performance and has been installed for years in Japanese process pumps. Based on these experiences and as long as proper workmanship and tools are employed, these cages absolutely qualify for process pumps.

However, they are not recommended for process pumps at facilities with lax installation tools or at plants that tolerate indifference toward the special needs of bearings with plastic cages. With plastic cages, one cannot allow working temperatures at assembly (or during repair) to exceed the temperatures at which certain polyamides tend to soften. Also, certain types of wear and progressive polyamide cage degradation are not being picked up by low-to-moderate cost means of vibration detection, usually employing a portable data collector-analyzer or data terminal. In contrast, degradation of machined brass cages shows up more readily on these portable devices during predictive maintenance assessments on process pumps.

Note the different number of ball pockets in the snap-in steel cage for

Figure 5.14: Snap-in pressed steel cages for Conrad-type bearings (left) and filling slot bearings (right).

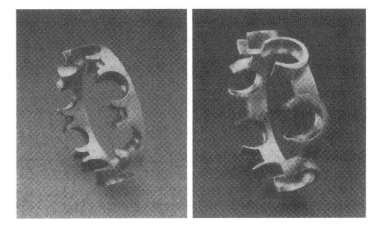

Fig, 5.15: Snap-in polyamide cage (left) and machined brass cage (right), both for Conrad-type bearings.

Conrad-type (deep groove) bearings (Figure 5.14, left) and a snap-in steel cage for filling slot bearings (Figure 5.14, right). Because an additional ball can be accommodated through the filling slots shown in Figure 5.16, such bearings will have radial load capacities approximately 10-15% greater than those of otherwise dimensionally equivalent Conrad bearings.

However, filling slots greatly diminish axial load capacity; also, bearing balls rolling over the edge of a filling slot may cause vibration or interruptions in the desired continuous oil film. Because of those risks and much adverse experience, filling slot bearings are not acceptable for process pumps in modern plants.

BEARING PRELOAD AND CLEARANCE EFFECTS

Bearing balls and raceways always deform slightly under load. Each bearing type or style or size has its own behavior when loaded. The result of deformation under load is easily visualized in an earlier illustration (Figure 5.13). Suppose loading the shaft axially from

Figure 5-16: Filling slot bearing (double-row style) with "stay-rod" cages.

left-to-right and looking at shaft movement relative to the housing end cap would result in the shaft end moving 0.001 inch (25 μ) to the right. In that case, the right-side row of bearing balls might no longer contact its inner ring raceway and would tend to skid. Skid situations wreck bearings and must be avoided, as explained in the earlier narrative dealing with upgrading and retrofit opportunities.

Preloading can avoid skidding and extend bearing life (Figure 5.17). Note that excessive clearance shortens bearing life. If the thrust loads acting on bearing "A" and bearing "B" are different and/or if skidding is to be avoided, then it might be best to select bearings with nonequal load angles. However, the load angles in Figure 5.18 are the same. Also, the plot shows a positive axial force P'. This indicates that the manufacturer of this particular matched set of bearings designed and manufactured each of the two inner rings a small amount narrower (perhaps 0.0002 inches or 5 μ) than the two outer rings. The two inner rings will make contact only after the castellated (cog-type) shaft nut has been tightened. When properly tightened, the 0.0004-inch (10-μ) clearance will have been reduced to face-to-face contact with no clearance. The preload P' exists before an operating load is superimposed.

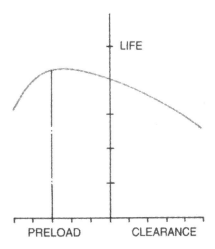

Figure 5.17: Slight preloading prevents skidding and slightly increases bearing life (by typically 15%). Operating with excessive preload or with bearing-internal looseness cause bearing life to decrease (Ref. 1).

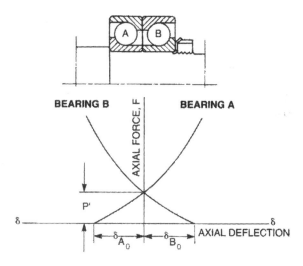

Figure 5.18: Axial deflection is a function of bearing geometry and force acting on a bearing. Note the castellated nut and tab washer, top half of illustration (Ref. 1).

Bearing Dimensions and Mounting Tolerances

There are important differences in bearing internal design clearances, manufacturing tolerances, and bearing mounting tolerances. Many of these are thoroughly explained in bearing manufacturers' tables, listings, and other literature. Internal clearances and other highly pertinent information is then coded as identification numbers and letters—"alphanumerics." Bearing numeric codes are standardized and contain four digits, (e.g., 7214). The first two digits refer to style, and 72 would indicate an angular contact ball bearing. Multiplying the second two digits by a factor of five gives the nominal bore dimension, in this instance $14 \times 5 = 70$ mm. On precision bearings, the four-digit number is always followed by suffixes.

Bearing alphanumerics without a suffix are a sure indication of inexpensive commodity bearings. Such bearings are cheap initially and their cost advantage is short-lived. They are prone to cause reliability issues in modern process pumps. The various suffixes generally differ among manufacturers, and some suffixes can be important. Not only would a complete listing of all available suffixes fill dozens of pages but these lists would have to be periodically updated.

As an example, the designation suffixes used in 2010 to identify certain features of SKF-brand, double-row angular contact ball bearings are explained in Table 5.1. From our earlier discussion, double-row ball bearings with the suffix "D" would require a threaded shaft end, bearings with the suffix "E" would be disallowed, the bearing fit should be C3 for process pumps, and so forth. This small example illustrates why one must pay attention to suffixes.

Again, whenever alphanumeric suffix information is disregarded by the pump user, failure risks increase and calamitous consequences become much more probable. Unless you have in-house expertise, consider linking up with the application engineering group of a highly knowledgeable bearing manufacturer.

There is no substitute for solid workmanship and conformance to the size and tolerance stipulations of competent bearing manufacturers. Only if there is no access to guidance from a competent manufacturer might one apply standard requirements to process pump bearings (in the 45- to 80-mm size ranges); if used in the location normally associated with the term radial, then such bearings should have shaft interference fits ranging from 0.0003 to 0.0007 inches (7 to 17 µ). Yet, because both shaft and bearing produc-

Table 5.1: Supplementary designations (suffixes) for SKF double-row angular contact ball bearings

A	No filling slots
CB	Controlled axial internal clearance
C2	Axial internal clearance smaller than Normal
C3	Axial internal clearance larger than Normal
D	Two-piece inner ring
E	Max type bearing, with filling slot
HT51	High-temperature grease for operating temperatures in the range −30 to +140 °C
J1	Pressed snap-type steel cage, ball centered
M	Machined window-type brass cage, ball centered
MA	Pronged machined brass cage, outer ring centered
MT33	Grease with lithium thickener of consistency 3 to the National Lubricating Grease Institute (NLGI) scale for a temperature range −30 to +120 °C (normal fill grade)
N	Snap ring groove in the outer ring
NR	Snap ring groove in the outer ring with snap ring
P5	Dimensional and running accuracy in accordance with ISO tolerance class 5
P6	Dimensional and running accuracy in accordance with ISO tolerance class 6
P62	P6 + C2
P63	P6 + C3
TN9	Injection molded snap-type cage of glass fiber reinforced polyamide 66, ball-centered
2RS1	Sheet steel reinforced contact seal of acrylonitrile-butadiene rubber (NBR) on both sides of the bearing
W64	Solid oil filling
2Z	Shield of pressed sheet steel on both sides of the bearing

ers make their respective products with manufacturing tolerances, situations might develop in which the shaft is near its high-tolerance limit and the bearing is at its low-tolerance limit. In that case, the resulting interference fit would greatly exceed 0.0007 inches and the bearing might be severely preloaded. Severe preloads create heat and reduce bearing life (Figure 5.17). To avoid excessive interference fits, the mating parts (shaft and bearing bore) each must be carefully measured at assembly. That kind of measuring takes time and only the true BOC pump users have institutionalized these measurement routines.

Paying close attention to bearing mounting and tolerance dimensions is even more critical on the thrust bearing side of process pumps. If the bearing manufacturer supplied back-to-back mounted angular contact bearings that are flush-ground and thus have no preload, then the mounting tolerance band would be the same as for radial bearings. If, however, these back-to-back bearings have a small gap between the inner rings and will thus become preloaded upon torque being applied to the shaft nut (see Figure 5.18), then an interference fit greater than 0.0003 inches (7-8 μ) should be avoided.

In back-to-back bearings with a small gap between the two inner rings, having greater than 0.0003 inches (7-8 μ) of shaft-to-bearing inner ring interference would put radial preload on top of axial preload. For such a bearing (i.e., the one with operating load superimposed on excessive preload) to survive it would require oil application method, lubricant properties, and other parameters near perfection. This near-perfection may be an unrealistic expectation in most situations.

Our generalized dimensional recommendations pertain to typical alloy steel shaft materials. It should be noted that certain stainless steels have higher coefficients of thermal expansion than AISI 4140 and similar alloy steel. For stainless steel shafts, the stipulated typical interference fits may have to be relaxed by a few percent. In all instances, outer ring installation and assembly tolerances should be loose fits ranging from 0.0002-0.0012 inches (5-30 μ). Again, it is worth recalling that the bearing manufacturer will produce outer ring outside diameters within a given tolerance band. Accurately measuring the difference between housing bore and bearing outside diameter may be difficult. Electronic or air-gauge measurements may be feasible, and plug gauges are another possible option. We are cautioning against entrusting bearing-related pump repairs to an unskilled work force.

These issues are just a small part of many "form tolerance issues" for bearings and shafts; form tolerances include perpendicularity of shoulders, concentricity, and out of roundness of cylindrical or tapered surfaces, and so on. But form tolerances are a complex subject that takes time to master. Selection and execution of the proper shaft and housing fit is just one part of making sure a bearing runs properly.

Shoulder run-out is typically held to a 0.0003 inch (7-8 μ) maximum, and shaft straightness deviations should typically not be allowed to exceed

0.001 inch (25 µ). Pump manufacturers and repair shops must adhere to seat perpendicularity (squareness) requirements. Returning to Figure 5.7, ascertain that there is no interference of the fillet radii at r_a and r_b. Also, be sure the D_a and d_a dimensions are selected to leave 25-35% of the adjacent inner and outer ring side face exposed. All are equally important considerations that mandate inspection and verification. Many shops do not have the training, skills, or equipment needed to produce quality shafts or housings, to repair in conformance with proper specifications.

However, making use of every bit of the bearing manufacturer's application engineering know-how and ascertaining that sound practices were applied throughout will result in long-lasting, reliable pumps.

WHAT WE HAVE LEARNED

1. Using long-term reliable bearing arrangements for process pumps has far-reaching effects. Important issues include components associated with the bearings, such as the shaft and housing.
2. Lubes and lubricant application (Chapter 7) interact and influence bearing success because they prevent wear and protect against corrosion.
3. Bearing protection (Chapter 8) is needed under both running and nonrunning (standstill) conditions. Oil cleanliness (Chapter 12) has a profound effect on bearing life. To design a long-term successful rolling bearing arrangement it is necessary to:
    *   Select the right size and type of bearing
    *   Determine the right fits, clearances, and preloads as well as to accommodate thermal expansion
    *   Apply lubricants properly and in the correct amount
    *   Be extremely mindful of pressure equalization needed to have uniform oil levels
    *   Closely examine housing geometries and ascertain that there are balance passageways that ensure equal pressures exist on each side of the radial and thrust bearing (see Figure 6.5, later, in which a bearing housing incorporates such passageways)
    *   Understand that bearing cage orientation (windage) and similarly overlooked issues could influence pump bearing life
4. Each design or service decision affects the performance, reliability, and economy of the bearing arrangement.

5. The amount of work involved in initial bearing selection or later remedial action depends on whether experience is already available in regard to similar arrangements. When experience is lacking, when extraordinary demands are made, or whenever failure history or life-cycle costs mandate, merely buying more replacement bearings makes little sense.

6. An acceptable range of "deviation from perfection" exists for each of the many bearing and lube-related parameters. Still, if operation at the limits allowable for more than one parameter is attempted, then bearings cannot possibly achieve maximum service life.

Finally, some pump models must be upgraded or modified to optimize bearing life. These optimization steps are often overlooked; they will be discussed in many chapters that make up this text.

References

1. SKF USA, Inc., Kulpsville, PA; Catalog 4000 US (1991).
2. American Petroleum Institute, Alexandria, VA; Standard API-610 (periodically updated by joint industry committees or task forces and issued as "latest edition").
3. SKF USA, Inc., Kulpsville, PA; Publication M190-730 "MRC Engineering Handbook," (1993).
4. SKF USA, Inc., Kulpsville, PA; Revised Publication M230-710, "MRC Bearing Solutions for the Hydrocarbon Processing Industry," (1996).
5. SKF USA, Inc., Kulpsville, PA; Original Publication M230-710, 3/91 ABG "MRC Bearings for Pumps," (1991).
6. MRC Bearing Services, Jamestown, NY; Publication M212-401, "MRC Extra-Wide Ball Bearing Retrofit Kit," (1991).

Chapter 6

Lubricant Application and Cooling Considerations

Bearing topics and lubricant application topics overlap in process pumps. The main issue here is that not all pumps are designed and sold with provisions ensuring that lubricant is consistently reaching the bearings. Many pumps will benefit from upgrading, and simply repairing them will not reduce failure risks. The fundamental root causes of failure must be discovered and remedied (Chapter 16).

The influence of windage on oil flow was discussed in Chapter 5; the influence of windage on oil flow illustrates the interdependence of bearing design and lube matters. Windage can be one of the overlooked root causes of bearing distress. It is considered a crossover topic and will be revisited in this chapter because it is too important to overlook.

LUBRICANT LEVEL AND OIL APPLICATION

Oil bath or sump lubrication is one of the oldest and simplest methods of oil lubrication; only grease lubrication is older than oil bath lube. The rolling elements pass (or "plough through") this oil sump during a portion of each shaft revolution (Figure 6.1). Oil bath lubrication is feasible unless and until too much frictional heat is generated by the plowing-through action of rolling elements

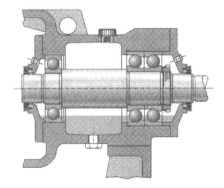

Figure 6.1: A typical pump bearing housing with oil level reaching to center of lowermost rolling elements. Here, keeping the DN-values below 6,000 reduces the risk of oil overheating.

at excessively high speeds. Because heat accelerates the rate at which oil oxidizes, the oil bath lube method is avoided on process pumps whenever DN, the inches of shaft diameter (D) multiplied by shaft revolutions-per-minute (N) exceeds 6,000.

To illustrate the DN approach, it can be reasonsed that a 2-inch shaft at 1,800 rpm, with a DN value of (2)(1,800) = 3,600, would operate in the suitable-for-oil-bath zone where no oil rings would be needed. Pumps incorporating a 2-inch shaft operating at 3,600 rpm (DN = 7,200) would use oil rings (shown earlier in Figure 5.3) to lift or spray the oil from a sump with its level maintained below the bearing. Less frictional heat results from lower oil levels (Figure 6.2) than from an oil level reaching the center of the bearing elements (Figure 6.1).

Issues with Oil Rings

A bearing housing with a lower oil level and intended for DN values in excess of 6,000 is shown in Figure 6.2. It will require the addition of either an oil ring or a similar flinger device to lift or spray-feed oil into the bearings.

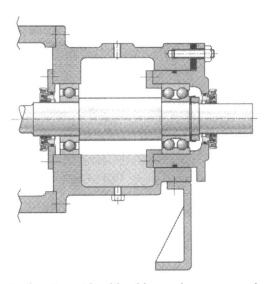

Figure 6.2: Bearing housing with oil level lowered to accommodate high DN values. Not shown: A flinger disc or other means of spraying (or lifting) oil into bearings. Also, oil pressure/temperature equalizing passages must be provided at each bearing (Ref. 1).

(See also Chapter 7 for comments on the interaction of oil rings and lubricants with different viscosities.)

Oil rings (also called slinger rings and shown earlier in Figure 5.3) can become progressively more unstable as DN values approach or exceed 8,000. Already, in the 1970s, a then prominent pump manufacturer wrote that its reliable pumps incorporated an "anti-friction oil thrower ensuring positive lubrication to eliminate the problems(!) associated with oil rings" (Ref. 1).

Oil rings are rarely, if ever, the most reliable lubricant application method. Oil rings often skip, misalign, and abrade (Figure 6.3). The shaft system must be near perfectly horizontal, and ring depth of immersion in the lubricant must be in the right range—usually close to 5/32 inch or 8-10 mm.

To avoid ring abrasion and dangerous oil contamination, ring eccentricity must be within 0.002 inches (0.05 mm) and surface finish should be reasonably close to 64 RMS. Oil viscosity should be in close range of typical International Standards Organization (ISO) VG 32 properties and temperatures must be in the moderate range (Ref. 2). The many different important parameters are rarely all within their respective desirable range in actual operating plants. If several individual parameters are just "borderline acceptable," then oil rings will intermittently malfunction.

Grooved oil rings perform slightly better than the plain or flat oil ring variety. Certain plastics perform a little better than the brass or bronze rings typically found in pumps (Ref. 4). It also has been argued that oil rings work better with ISO Grade 46 oil compared with ISO Grade 68 oil. (A well-formulated synthetic Grade 32 may be required to suit both the constraints of oil rings and the needs of a particular pump bearing; Chapter 7 has more details on the issue.)

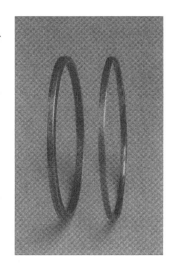

Figure 6.3: Oil rings in as-new (wide and chamfered) condition on left and abraded (narrow and chamfer-less) condition on right (Source: Ref. 3).

Because oil ring behavior is difficult to control, some reliability-focused purchasers try to avoid them. These users often specify and select pumps with large diameter flinger discs (Figure 6.4) in DN applications in which lower oil levels are needed. *Small* diameter flinger discs are some-

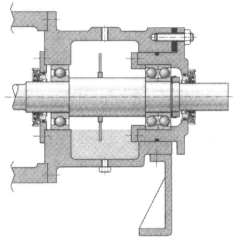

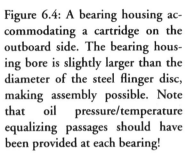

Figure 6.4: A bearing housing accommodating a cartridge on the
outboard side. The bearing housing bore is slightly larger than the
diameter of the steel flinger disc,
making assembly possible. Note
that   oil   pressure/temperature
equalizing passages should have
been provided at each bearing!

times used in oil bath applications (Figure 6.5) simply to prevent temperature stratification of the oil. Without them, hot oil would tend to float at
the top.

PRESSURE AND TEMPERATURE
BALANCE IN BEARING HOUSINGS

Note that the double-row radial bearing in Figure 6.5 was—in the
1960s—called a "line bearing," and take a careful look at the balance
holes in the bearing housing near the top of each of the two bearings.
Then-prominent manufacturer Worthington Pump Company knew the
importance of achieving pressure balance throughout the entire pump
bearing housing. Oil leakage risk past the oil seals (lip seals in this old design) was greatly reduced by keeping all pressures equal. Pressure balance
also is important because it largely neutralizes the potential windage effects of certain slanted bearing cage configurations or angularly inclined
orientations.

Flexible flinger discs are sometimes used to enable insertion in the
simplest bearing housings. In these, the bearing housing bore diameter is
smaller than the flinger disc diameter. To accommodate the preferred *solid*
steel flinger discs, bearings must be cartridge-mounted (Figures 6.2 and 6.4).

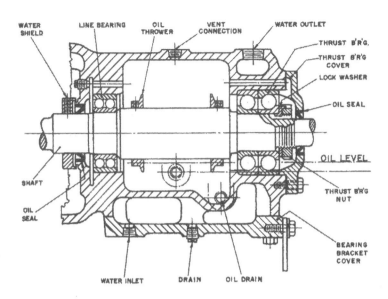

Figure 6.5: This 1960s vintage bearing housing uses "oil throwers" for keeping the oil at uniform temperature. Note pressure balance holes provided at the top of radial and thrust bearings (Source: Worthington Pump Installation and Operating Manual, ~1966).

With a cartridge design, the effective bearing housing bore (i.e., the cartridge diameter) will be large enough to allow insertion of a steel flinger disc with the needed diameter. Providing such a cartridge will add to the cost of a pump. In the overwhelming majority of cases, the incremental cost of the cartridge design will be low compared with what it costs to repair a pump.

Again, neither Figure 6.2 nor Figure 6.4 shows balance holes or passageways that would ensure pressure and temperature equalization. Internal pressure and temperature balance between the central volume of the bearing housing and the spaces between bearings and end caps is essential. This requirement seems to be disregarded in some pump models.

As shown in Figure 6.5 and earlier in Figure 5.5, drilled or similarly machined passageways are needed near either the top or the bottom of the bearing housing bore to obtain pressure and temperature equalization. In some cases, it would be best to provide equalization passages at both top and bottom. Lack of these passages is one of several explanations for oil leakage and overheated oil. Overheated oil and/or oil contaminated with slivers of O-ring material will result in "black oil."

Note also that the 1960s vintage bearing housing of Figure 6.5 shows lip seals and water cooling provisions. Cooling water was deleted from pumps with rolling element bearings in the early 1970s (Ref. 5). The lip seals shown here would no longer be acceptable, and modern bearing protector seals would be used instead. (See Chapter 8 for details.)

COOLING NOT NEEDED ON PUMPS WITH ROLLING ELEMENT BEARINGS

Cooling is still found in many pumps with high speed and heavily loaded rolling element bearings. But cooling of the oil is rarely needed and often of no benefit in installations with rolling element bearings. Suppose a cooling jacket restricts the bearing outer ring from free thermal growth in all radial directions. However, the bearing inner ring heats up and grows, causing bearing internal clearances to vanish. An excessive preload could result. (Chapter 5 dealt with bearing preloads in more detail.)

Similarly, immersing cooling coils in the oil will cool not only the oil but also the air in the bearing housing. Such cooling then tends to promote moisture condensation and harmful oil contamination. Therefore, cooling water has been deleted from every pump with rolling element bearings at many best-of-class (BOC) locations (Refs. 1 and 5). Since the late 1970s, there no longer is cooling water on bearing housings of pumps with operating (fluid) temperatures up to and including 740°F (394°C) in modern BOC oil refineries.

Because cooling water ports are shown on a pump drawing, the user is led to believe that such cooling is either needed or helpful. Commenting again on Figure 6.5, when it was discovered that cooling is no longer needed, BOC companies began to leave these cooling water drains open. With modern synthetic lubricants and properly selected rolling element bearings, cooling is no longer used in process pumps. So, irrespective of the lube application method, on *rolling element* bearings cooling will not be needed as long as high-performance synthetic lubricants are used.

With sleeve bearings, cooling is used to maintain optimum oil viscosity through close temperature control. This provides a reasonably stable environment for oil rings in pumps equipped with *sleeve bearings*. Circulating systems are the primary choice for large pumps that incorporate these bearings.

Large process pumps use sleeve bearings and circulating oil systems. In circulating systems, the oil can be passed through a heat exchanger before being returned to the bearing. Pressurization is needed to move oil through filters and exchangers, but the bearing itself is rarely fully pressurized. In some sleeve bearing systems, oil rings are used to lift the oil and deposit it on the shaft surface. In other sleeve bearings, an oil spray from suitably placed nozzles is directed to oil grooves with good effect and high reliability (Refs.1 and 6).

OIL DELIVERY BY CONSTANT LEVEL LUBRICATORS

Since the late 1800s, various types of constant level lubricators have been used successfully. Unfortunately, the vulnerabilities of some constant level lubricators are often overlooked. If there are pressure differences between regions in the bearing housings and regions in the lubricator, then the two oil levels will differ. Also, whenever caulking is used to secure an oiler bulb or transparent bottle to a supporting component, the caulking will ultimately lose its resiliency or elasticity. Small fissures will develop and rainwater will enter at these fissure locations via capillary action. Therefore, constant level lubricators must be part of conscientious preventive maintenance action. They must be included in a precautionary replacement strategy.

Not all versions of "constant level" lubricators supplied by pump manufacturers will best serve the reliability-focused user. In the widely used pressure *nonbalanced* devices equal or similar to Figure 6.6, the oil level below the reservoir bottle (at the large wing nut that supports the oiler bottle) is contacted by ambient air. But that is so only as long as the pressure in the pump bearing housing to which this device is connected is also at ambient pressure. Should the pressure in the connected pump bearing housing be above ambient, the oil level in the bearing housing would be pushed down and the oil level in the constant level lubricator base (at the tip of the wing nut) would rise.

Look closely at the bearing in Figure 6.6(a) and note how a small decrease in oil level might stop lubricant from flowing into the bearing. Observe also how the needed mounting location is indicated by the clockwise rotational arrow in Figure 6.6 (a).

In the *pressure-balanced* device of Figure 6.7, level differences are far less likely to occur. Here, the space surrounding the slanted tube is always at the same pressure as the vapor space in the bearing housing.

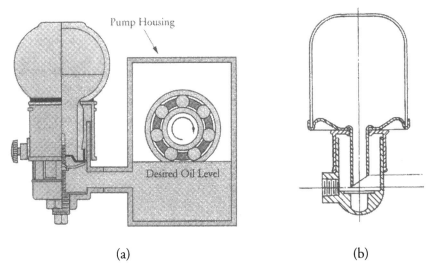

(a)   (b)

Figure 6.6: Two different constant level lubricator styles, both are pressure nonbalanced. In (a), the wing nut threaded into the vertical rod establishes the oil level. In (b), the upper horizontal line represents the oil level. Note that ambient air pressure exists above that line in (b) and above the wing nut in (a). (Source: Ref. 3).

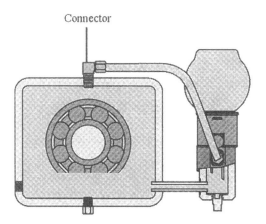

Figure 6.7: Pressure-balanced constant level lubricator (Source: Ref. 3).

BLACK OIL

Refer back to Figure 6.6(a). If this oil level is decreased to a point much below the center of the lowermost rolling element, then the risk of oil over-heating increases drastically. Many oil formulations typically used in process pumps turn black when overheated. This black oil then shows up in the transparent bowl of constant level lubricators and does so because thermal convection currents cause oil to circulate, and because temperature-dependent cyclic compression and expansion occur in the air space at the top of constant level lubricator bowls. Black oil can show up in the bowl because of convection.

In situations in which the oil reaches the center of the lower-most bearing element, there will be an increase in oil temperature because the rolling elements encounter friction as they move through the lubricant. This temperature increase tends to be excessive at high DN values—another possibility for black oil formation.

Black oil at pump start-up is sometimes the result of an oil ring being wedged into the space between the shaft surface and a long screw that is supposed to limit oil ring travel (shown earlier in Figure 2.1).

It is also possible that inadequate lube oil formulations lack the film strength to maintain a separating oil film between rolling elements and brass or bronze cages at initial pump start-up. Although brass or bronze cages are recommended by API-610 for centrifugal pumps, this particular vulnerability is thought to be greater with copper-containing cage materials than with ferrous metals and suitable plastics.

When there is no longer an adequate oil film, considerable heat will be generated and small portions of the cage material will transfer to the rolling elements. The oil will discolor and contamination can range from minimal to severe.

Additional comments on black oil can be found in Chapter 8, dealing with bearing protector seals. It will again explain that black oil is either overheated oil or lubricant contaminated with slivers of abraded O-ring material.

LUBRICANT APPLICATION AS OIL MIST (OIL FOG)

Plant-wide oil mist lubrication systems have proven superior to other methods of lubricating rolling element bearings in process pumps. These systems can supply oil in the form of a dense mist to more than 100 pumps and are limited only by distance traveled (Ref. 7). Small oil mist units are available for serving as lube application modules on up to four pumps (one was shown in Figure 3.3). Used in conjunction with pump and motor bearing housings configured per Figure 6.8, closed oil mist systems will not emit stray mist fog into the environment.

Oil mist has the appearance of dense fog; it usually consists of a 200,000:1 (by volume) mixture of dry instrument air and atomized lubricating oil (ISO VG 68 or 100). The mist is produced in a simple mixing nozzle and then travels through the plant at a pressure of 20-35 inches (500-890 mm) of H_2O. The header system is unheated, noninsulated, and at ambient temperature. Individual branch lines come off the top of a header pipe and route the mist through a small metering orifice ("reclassifier") to the rolling element bearings of pumps and their electric motor drivers (Figure 6.8).

Because no oil level exists in the oil sump, the pure oil mist method of lubrication is also called dry sump oil mist. With no liquid oil to "lift" from a sump, oil rings or flinger discs no longer exist.

Electric motor drivers and entire non-running (standby) pump sets are included in plant-wide oil mist systems. Dry sump (pure) oil mist applied per Figure 6.8 clearly eliminates every one of the previously discussed prob-

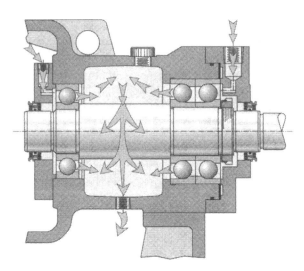

Figure 6.8: Oil mist lubrication applied to a pump bearing housing in accordance with API-610, 10th Edition. Note dual mist injection points and use of face-type bearing protector seals to prevent mist from escaping to the atmosphere (Ref. 7).

lems with oil rings. It also avoids or eliminates as well as supersedes issues with defective constant level lubricators and missing balance holes in bearing housings (Ref. 7).

Cost justification calculations show payback periods typically in the range from 6 months to 3 years for many oil mist installations. The more rapid payback is associated with, among other things, fewer bearing failures resulting in a reduced number of fire incidents in process plants (Ref. 1). The less rapid payback is for environmentally friendly "closed" oil mist systems.

All things considered and for process plant applications, dry sump (pure) oil mist is by far the most reliable, predictable, and least risk means of providing trouble-free bearing lubrication for process pumps. Oil rings are no longer present, and black oil is not experienced with oil mist applied per Figure 6.8. Note that face-type bearing protector seals are used in up-to-date oil mist lubricated bearing housings. The through-flow routing in Figure 6.8 complies with the requirements stipulated in the year 2000 and later editions of API-610 (Ref. 7).

The reliability and availability of modern oil mist systems exceeds other methods of pump lubrication. At one plant and over a period of 14 years, a single qualified contract worker serviced the 17 systems by visiting the facility one day each month. In this 14-year time period, there was only one malfunction; it involved a defective float switch in the bulk oil container of one of the 17 oil mist generator consoles. The incident caused a string of pumps to run without a functioning mist supply system for 8 hours. Still, there were no bearing failures. In this plant the combined long-term availability and reliability of the oil mist systems was calculated to have reached 99.99962% (Ref. 8).

The advantages and disadvantages of oil mist lubrication as compared with traditional liquid oil sump (oil bath) lubrication are summarized as follows:

Advantages:
- Represents automated lube application for pump, driver, and standby set
- Reduces bearing failures by 80-90%
- Lowers bearing operating temperatures by typically 10-20°F (~6-12°C)
- Continuously removes bearing wear particles
- Slight positive system pressure eliminates contaminant entry
- Reduces energy costs by 3-5%

- Reduces oil consumption by about 40%
- Greatly reduced maintenance intensity because oil mist generator and its associated equipment contain no moving parts

Disadvantages:
- Capital investment
- Cost of compressed air

DESICCANT BREATHERS AND EXPANSION CHAMBERS

Oftentimes, desiccant-breather combinations seem to address the *symptom* of a housing-internal, pressure-related shortcoming and might even represent a fix, in some isolated instances. No such breathers are needed in closed bearing housings or if pressures are equal in front and behind bearings inside of pump bearing housings.

If used, then desiccant breathers are maintenance items that must be considered in the budget. These elements or containers are filled with a chemical that absorbs moisture. The container is often made of a transparent polycarbonate or similar plastic material.

Some users also believe in expansion chambers (Figure 6.9) and have occasionally modified the tops of pump bearing housings to allow installation of both a desiccant breather and an expansion chamber. Expansion chambers serve no technically advantageous purpose in vented pump bearing housings.

In fully sealed bearing housings the pressure increase is a function of temperature increase. The degree to which bearing housing pressures are reduced by expansion chambers is then a function of the volumetric ratio of the chamber and the bearing housing volume. (Recall the combined gas laws: $P_1V_1/T_1 = P_2V_2/T_2$.)

If a pump has a relatively large bearing housing volume and/or a

Figure 6.9: Expansion chamber intended for process pump bearing housings (Ref. 3).

low temperature increase, then adding a small expansion chamber will serve little purpose. Applying the gas law equation will prove it.

WHAT WE HAVE LEARNED

There are truly hundreds of combinations of oil application, bearing lubricant bypass possibilities, pressure equalization options, and venting arrangements. Examining a bearing housing cross-sectional drawing is the minimum requirement for a competent failure analyst. Physically examining a bearing housing and its bearings is even better. (Remember: you get what you inspect, not what you expect).

The following is a recap relating to shaft diameter and shaft rpm interaction:

- Sump-lubricated pumps with DN [dn] values (shaft diameter in inches [mm] times rpm) up to 6,000 [dn~160,000] allow lube oil to reach the center of the rolling elements at the 6 o'clock position.
- Total flooding of pump bearings (regardless of DN or dn value) may result in excessive heat generation.
- For pumps with DN [dn] values greater than 6,000 [dn~160,000], the oil level may have to be lowered and oil rings or flinger discs may have to be chosen by the user or pump manufacturer.
- Many oil rings are unstable and skip at DN [dn] values greater than 6,000 [dn~160,000]; some best-of-class companies will not use them above DN [dn] values greater than 8,000 [dn~213,000].

Among the elusive reasons why pump bearings fail, we will find the following:

- An occasional unreliable wet sump oil mist application
- Dry sump oil mist introduced in a nonoptimized manner (i.e., arrangements that do not conform to the 10th Edition of API-610, see Ref. 9)
- Unexpected pressure drops through desiccant containers
- Misunderstandings as to what a bearing housing expansion chamber will and will not do

We learned to pay attention to avoiding pressure gradients in the vicinity of bearings. The bearing cages of Figure 5.2(e) act as impellers and create windage effects that make pressure balance within the bearing housing difficult.

Next to the vastly superior dry sump oil mist, installing a bulls-eye sight glass and omitting the constant level lubricator is the closest acceptable alternative. A properly dimensioned flinger disc and modern bearing protector seals (Chapter 8) would be part of this closest acceptable alternative.

An inexpensive "all purpose" or "standardized" lubricant will give mediocre results at best. In the end, buying the right lubricant formulation—even at a premium price—will be the most cost-effective option for reliably lubricating process pumps (Ref. 10).

Many different variables influence oil levels in pump bearing housings, windage and venting are among them. No statistics are available that allow definitive cataloging or linking of the many factors, but doing all things right is the path to satisfactory pump operation. Although a small deviation from the norm can sometimes be tolerated, allowing several parameters to reach their individual limits of acceptability will always have a negative effect on pump reliability.

References

1. Bloch, Heinz P., and Alan Budris; *Pump User's Handbook: Life Extension*, 2nd Edition (2006), The Fairmont Press, Lilburn, GA. (ISBN 0-88173-627-9).

2. SKF USA, Inc., Publication 100-955, Second Edition, "Bearings in Centrifugal Pumps," (1995).

3. TRICO Manufacturing Corporation, Pewaukee, WI; Commercial Literature and www.tricocorp.com.

4. Bradshaw, Simon; "Investigations into the contamination of lubricating oils in rolling element pump bearing assemblies," Proceedings of the 17th International Pump User's Symposium, Texas A&M University, Houston, TX, 2000.

5. Bloch, Heinz P.; *Machinery Reliability Improvement*, 3rd Edition (1998), Gulf Publishing Company, Houston, TX. (ISBN 0-88415-661-3).

6. Bloch, Heinz P.; "Inductive pumps solve difficult lubrication problems," *Hydrocarbon Processing*, September 2001.

7. Bloch, Heinz P., and Abdus Shamim; *Oil Mist Lubrication—Practical Application* (1998), The Fairmont Press, Lilburn, GA. (ISBN 088173-256-7).

8. Bloch, Heinz P., and Fred Geitner; *Machinery Component Maintenance and Repair*, 3rd Edition (2005), Gulf Publishing Company, Houston, TX. (ISBN 0-7506-7726-0).

9. American Petroleum Institute, Alexandria, VA; API-610, "Centrifugal Pumps," 10th Edition, (2009).

10. Bloch, Heinz P.; *Practical Lubrication for Industrial Facilities*, 2nd Edition (2009), The Fairmont Press, Lilburn, GA. (ISBN 088173-579-5).

Chapter 7

Lubricant Types and Key Properties

LUBRICANT VISCOSITIES

Viscosity is by far the most important property of the lubricants applied to process pump bearings, and practical texts deal with the issue in great detail (Ref. 1). In general, thicker oil films will give better pump bearing protection than thinner oil films. For process pumps with rolling element bearings, International Standards Organization (ISO) Grade 68 lube oils will allow higher operating loads and they are generally preferred over ISO Grade 32 lubricants.

Not all application methods are possible with ISO viscosity grades higher than VG 32. How to capture the benefits of thicker lubricants without actually using them or how to apply ISO VG 68 and thicker oils best will be the subject of discussion later in the chapter.

Figures 7.1 and 7.2 allow quick determination if a particular lubricant selection is in the right range. Suppose we were dealing with a 1,800 rpm pump, and its bearings had bore diameters of 65 mm (dimension d) with bearing outer diameters (D) of 120 mm. Because $d_m = 0.5$ (d+D), in this case $d_m = 92$ mm. Enter the horizontal x-axis of Figure 7.1 at this calculated mean dimension and plot upward to intersect an imaginary 1,800 rpm diagonal line—just above the 2,000 rpm diagonal. From there, move to the left and read off ~11 mm²/s (a measure of viscosity that is more commonly called 11 centistokes, generally abbreviated as cSt).

We have now established that, in this example, the minimum kinematic viscosity required to give adequate protection at operating temperature is 11 cSt. Because oils become thinner when heated and if our operating temperature is high, we realize we should have selected thicker oil. As this thicker oil then reaches a higher operating temperature, it will, hopefully, not "thin out" to a viscosity below 11 cSt.

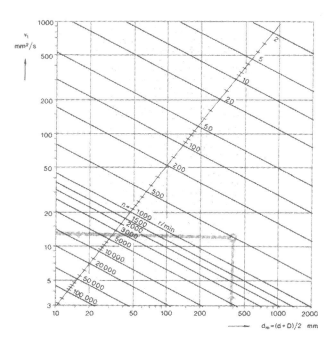

Figure 7.1: The required minimum (rated) viscosity (v_1) depends on shaft rpm and bearing size (Ref. 2).

Continuing on Figure 7.2, we must now either assume a certain bearing operating temperature, say, 70°C or 158°F. Entering the horizontal x-axis scale at 70°C and moving upward to ISO VG 68 would allow us to read off 20 cSt on the vertical scale—close to twice what we need. (Important: We should always ascertain that our oil delivery system works with this thicker-than-needed oil.) We might pick ISO VG 32 and, as long as our actual bearing operating temperature does not exceed our assumed temperature of 70°C, we would now read off (from the vertical y-scale) an operating viscosity of 11 cSt—just right.

We could also use Figure 7.2 by entering the vertical y-scale at the required 11 cSt and, after intersecting a particular viscosity grade oil, plot down to read off the maximum allowable bearing operating temperature on the horizontal x-scale. Thus, intersecting at, say, VG 68, would tell us that this oil could be allowed to reach 92°C and still satisfy our viscosity requirement of 11 cSt.

Thicker film oils are easily and reliably applied with the oil mist meth-

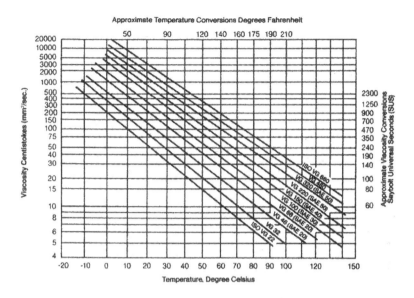

Figure 7.2: For a required viscosity (vertical scale), the permissible bearing operating temperatures (horizontal scale) increases as thicker oils are chosen (diagonal lines).

od briefly summarized in Chapter 6. However, thick oils can be difficult to apply with only the oil rings shown earlier in Figure 5.3.

Running a pump, one might ultimately achieve an operating temperature that allows a certain thick oil to flow nicely. But what if, at startup, the initial operating temperature is low and the oil will not flow? That is what often happens when someone buys a standard "*multipurpose oil*" without really thinking things through.

Recall or refer back to Figures 5.5 and 5.18, which showed lube applications with the oil level reaching to the center of bearing elements at the 6-o'clock position. In these instances, ISO Viscosity Grade 68 allows operation in a relatively wide range of ambient temperatures.

However, in applications using oil rings and with oil levels reduced below the periphery of the lowermost bearing component (see Chapter 5, Figures 5.3 and 5.19), ISO Grade 32 lubricants may have to be used simply because not all oil rings will work with thicker oil viscosities. At the DN values typically encountered in process pumps, most oil rings function better in ISO Grade 32 lubricants than they would in ISO Grade 68 oils.

The following are a few general guidelines worth considering:

- Using a mineral oil would generally require oil to be changed every 6-12 months. With a clean, premium-grade synthetic lubricant, one would typically extend oil change intervals to about 24 months.

- ISO Grade 32 *mineral* oils are often considered too thin for pump bearings. They rarely qualify for long-term, risk-free use in pumps equipped with rolling element bearings in typical North American and European ambient conditions. But simply switching to ISO Grade 68 *mineral oils* will be risky for bearings that are fed by oil rings.

- Appropriately formulated with the right base stock and with proprietary additives, ISO VG 32 *synthetics* are acceptable from film strength and film thickness points of view (Refs. 1 and 3). In fact, the performance of some ISO VG 32 synthetics duplicates that of ISO VG 68 *mineral* oils. These superior ISO VG 32 synthetics excel by simultaneously satisfying the requirements of sleeve bearings and rolling element bearings (Ref. 4).

- Superior synthetics achieve high film strength through proprietary additives. So, there can be significant differences in the performance of two lubricants of the same viscosity and using the same base stocks. Only one might be suitable for highest reliability services.

- The notion that one oil type or viscosity suits all applications is rarely correct and is easy to disprove. Similarly, no fixed or particular oil ring geometry is ideally suited for all oil types and viscosities. Custom-designed oil rings may be required to work with the thicker oils at certain high shaft DN values.

WHEN AND WHY HIGH FILM STRENGTH SYNTHETIC LUBRICANTS ARE USED

Quality lubrication includes sound and risk-free application method, proper lube quantity, appropriate oil type and viscosity, properly storing and handling the lubricant, attending to bearing housing contamination issues, and implementing appropriate oil change intervals.

To summarize good lubrication practices: we must choose the right oil, take proper care of it, and change it before bearings are harmed. Improvements in lubricant quality can only be achieved by using oils with superior lubricating properties. These would be premium synthetics.

Even among prominent synthetic lubricants, oil performance can vary greatly based on the amount and composition of additives in the oil. For process pump bearing lubrication, at least one company combines synthetic base oils including poly-alpha-olefin (PAO) and dibasic ester base stocks with advanced additive chemistry to realize greater film strength (Ref. 4).

Numerous incidents have been documented in which advanced lubrication technology has significantly improved pump reliability. In most cases, advanced lube technology with its often more favorable (lower) coefficient of friction results in reduced bearing operating temperatures. Micro-cracks in bearing surfaces cause increased noise and vibration (Ref. 5); suitable high film strength oils will fill these microcracks. This then lowers noise intensity and reduces vibration severity (Refs. 1 and 3).

High film-strength lubricants also lessen the probability of lube oil darkening during the running-in period of bearings with brass or bronze cages. There have been reported instances of high frictional contact during the initial run-in period of the copper-containing material recommended—for its other qualities—by API-610. If the net axial thrust action on one of the two back-to-back oriented bearings causes it to become unloaded, then it may skid (see Figure 5.8).

The risk of lube oils darkening during the run-in period of such pumps is reduced through the use of high film-strength synthetic lubes. To be fair, this risk could also be reduced by insisting on impeccable installation techniques and the selection of bearings with cages made of advanced high-performance polymers (see comment in Chapter 5).

Whatever the differential cost of a quart (or liter) of high film-strength synthetic, it is insignificant compared with the value of an avoided failure incident on critical, nonspared refinery pumps.

Critically important pumps, pumps in high-temperature service, and pumps that have failed more often than others in the plant's pump population should, therefore, be lubricated with high film-strength synthetic oils.

On pumps in which a problem is in progress, changing to a superior synthetic is highly recommended. If access to the sump drain is safe while the pump is in service, then the present oil can be drained while such pumps are online and running. Many superior synthetics are compatible with the oil presently used in a particular pump (Ref. 6).

Switching to superior film strength synthetic lubricants would give immediate payback. Virtually every cost justification calculation indicates unusually large benefits for employing these lubes on problem pumps.

Of course, there are certain pump bearing or lube degradation problems that have nothing to do with the lubricant type. In those instances, nothing will be gained by changing over to better oils. (There is never a substitute for responsible and accurate failure analysis, see Chapter 16.)

LUBRICANTS FOR OIL MIST SYSTEMS

Pure oil mist lubrication (Chapter 6) eliminates the need for either oil rings or flinger discs. No liquid oil sumps are maintained in the bearing housings; hence, the term "dry sump" is often used to describe modern oil mist lubrication. ISO VG 68 and VG 100 mineral or synthetic oils are used, although properly formulated ISO VG 32 *synthetics* (but not mineral oils) will serve most pump bearings as well as virtually all types of rolling element bearings in electric motors. Decades of experience on thousands of pumps and electric motors attest to the viability and cost effectiveness of modern plant-wide oil mist systems. Typical payback periods when using oil mist on problem pumps have generally been less than 1 year.

As shown in the bar chart of Figure 7.3, bearing friction can be reduced by switching to different oils, by going with a different lube application method, or by switching both lube application method and oil type. Five different modifications were closely examined in a cooperative effort involving a multinational lube oil producer and prominent bearing manufacturer. The results are plotted and percentage reductions in bearing friction displayed on the vertical scale of Figure 7.3.

WHAT WE HAVE LEARNED

- If repeat failures occur in process pumps, then there are overwhelming odds of several small deviations combining. Once several (in themselves tolerable) deviations combine, just one more blip becomes the proverbial "straw that breaks the camel's back." In other words, once several deviations exist, one additional deviation will often cause a serious failure.

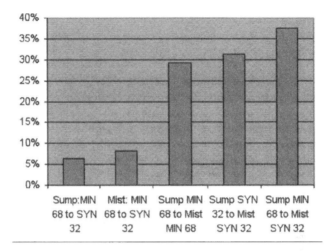

Figure 7.3: How changes in lube application, oil types and lube viscosities tend to affect percentage reductions in bearing friction; these are displayed on the vertical scale. (Source: Ref. 6).

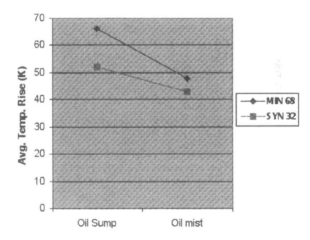

Figure 7.4: How changes in lube application, oil types, and lube viscosities tend to affect bearing temperature increase; these increases/decreases are displayed on the vertical "Kelvin" (°C) scale. (Source: Refs. 3 and 7).

- Using a "normal" lubricant is fine as long there is no excessive axial load on the bearings. With a high axial load, superior quality oils will be mandatory.

- Having an oil ring with an eccentricity slightly greater than the normally allowed 0.002 inches (0.05 mm) *seems* acceptable but only until perhaps the out-of-horizontality of the shaft system exceeds a certain value, or until there is—additionally—a certain deviation from a normally close range of lube oil viscosity, or until the increases in generally tolerable cavitation, bearing-related vibration, or whatever else converge to cause calamity.

Make it your business to understand these facts and act on this understanding. You do have it in your power to avoid process pump failures.

One has to be consistently inside the acceptable ranges of dimensional, material composition, fabrication-specific, application-related as well as a host of other parameters, disciplines, and ingredients. Adherence to sound specifications is not difficult once a proper mindset is cultivated. The difficulty is in cultivating the mindset.

References

1. Bloch, Heinz P.; *Practical Lubrication for Industrial Facilities*, 2nd Edition. (2009). The Fairmont Press, Lilburn, GA. (ISBN 0-88173-579-5).
2. SKF USA, Kulpsville, PA; General Catalog, Engineering Section. 2008.
3. Bloch, Heinz P. and Allan Budris; *Pump User's Handbook: Life Extension*, 3rd Edition (2009), The Fairmont Press, Lilburn, GA. (ISBN 0-88173-627-9).
4. Morrison, Frank R., Zielinsky, James, and James, Ralph; "Effects of Synthetic Fluids on Ball Bearing Performance," ASME Publication 80-Pet-3, February, 1980.
5. Wilcock, Donald F., and Richard E. Booser; *Bearing Design and Application*, (1957) McGraw-Hill, New York, NY.
6. Royal Purple Ltd., Porter, Texas, Marketing Literature, 2000-2004.
7. Villavicencio, E.; "Basic elements of tribo-thermo-dynamics," presented at the First International Symposium on Reliability and Energy Tribo- Efficiency, Mexico City, MEXICO, June 2002.

Chapter 8

Bearing Housing Protection and Cost Justification

Both oil and air fill the bearing housing. As either gets warm and expands, or cools and contracts, an interchange of the internal air takes place with the surrounding or ambient air. The interchange is called "breathing."

To stop this breathing and resulting contamination, the breather vents on the housings shown earlier in Figure 2.1 should be plugged and suitable bearing protector seals* installed where the shaft protrudes through the housing.

Inexpensive lip seals are sometimes used for sealing at the bearing housing, but lip seals typically last only about 2,000 operating hours—3 months (Ref. 1). When lip seals are too tight, they tend to cause shaft wear (Figure 8.1, top portion) and, in some instances, lubricant discoloration ("black oil," Ref. 2) and contamination. Once lip seals have worn and no longer seal tightly, oil is lost through leakage. The API-610 standard for process pumps disallows lip seals and calls for either rotating labyrinth-style or contacting face seals.

Figure 8.1: Lip seals (left and top right) tend to wear and have typically only a 3-month operating life; replacement with a high-reliability rotating labyrinth seal (below, right) is possible without first having to repair a damaged shaft surface.

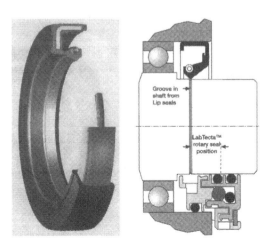

---

*The terms bearing housing protector seal, bearing protector seal, and bearing isolator are used interchangeably.

The bearing housing protector seal in the lower right portion of Figure 8.1 incorporates a small and a large diameter dynamic O-ring. This bearing protector seal is highly stable and not likely to wobble on the shaft; it is also field-repairable. With sufficient shaft rotational speed, one of the rotating ("dynamic") O-rings is flung outward and away from the larger O-ring. The larger cross-section O-ring is then free to move axially and a microgap opens (Ref. 2).

When the pump is stopped, the outer of the two dynamic O-rings will move back to its stand-still position. At stand-still, the outer O-ring contracts and touches the larger cross-section O-ring. In the purposeful design of Figure 8.1, the larger cross-section O-ring touches a relatively large contoured area. Because contact pressure = force/area, a good designs aims for low pressure. A good design (Figure 8.1) will differ greatly from the model shown in Figure 8.2(a), where contact with the sharp edges of an O-ring groove will cause O-ring damage.

Rotating bearing housing protector seals ("bearing protectors") are classed into the main categories contacting and noncontacting. Contacting styles include (a) plain lip seals, (b) lip seals engaging a tapered groove labyrinth, and (c) the various face-contacting types. Low-cost noncontacting original equipment manufacturer (OEM) bearing protectors are often of the simplest labyrinth configuration (Figures 8.2(a) and 8.2(b). A least-risk configuration is shown in Figure 8.2(c).

NONCONTACTING BEARING PROTECTOR SEALS

To recap: By definition, noncontacting implies at-speed operation with a small gap between rotating and stationary sealing components. Optimum gaps are found in advanced rotating labyrinth seals (Figure 8.2(c)).

In noncontacting bearing protector seals, the rotating component typically has a complex outer profile. This profile is located adjacent and in close radial and axial proximity to the complex inner profile of a stationary component.

Together and in theory, these complex profiles form a tortuous path that prevents inward or outward passage of unwanted materials or fluids. Protector seals are more likely to fail if the design places a dynamic O-ring close to razor-sharp circumferential edges or grooves (Figure 8.2(a)).

Note that bearing protector seals often differ in basic design and detailed geometry. The details may be important, (see Figure 8.3). Here, the re-

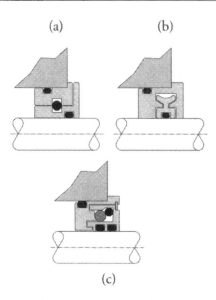

Figure 8.2: Noncontacting bearing protector seals: (a) plain labyrinth—its O-ring risks being damaged by sharp grooves; (b) sealing lip engaging a tapered groove, a high drag effect is possible; (c) dual dynamic O-ring and dual clamping rings as found in a highly reliable, field-repairable design.

liability professional must consider the effects of overhung mass and a single O-ring used for clamping. Rotor tilt will cause the squeezed clamping ring to become "kneaded" and the rotor to walk.

Figure 8.3 thus points to an elusive root cause of failure. In this instance, O-ring debris can end up in the bearing housing, causing the well-documented appearance of "black oil." (Ref. 3 and several other chapters in this book mention "black oil"—please see index.)

CONTACTING BEARING PROTECTOR SEALS

In the second category, contacting bearing protectors, users have also employed technically advanced hybrid seals (Figure 8.4) and face-contacting magnetic seals (Figure 8.5). In magnetic seals, the closing force is supplied by several floating permanent rare-earth magnets that exert a pulling force on steel Z-rings. Mounted in these Z-rings are carbon faces that then contact an O-ring mounted single piece rotor. Face-contacting seals are needed in pump

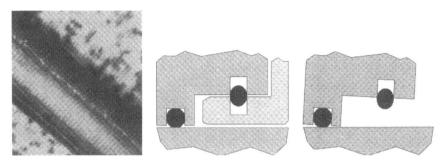

Figure 8.3: Inadequate shaft clamping, off-center mass, and O-ring contact with sharp edges can cause degradation and unforeseen contamination of lube oil.

bearing housings that receive lubrication from closed-loop oil mist systems (Ref. 4).

Both configuration-related categories of bearing protectors are available as designs that can accommodate angular shaft-to-housing misalignment (Figure 8.6). Variants for steam turbines have also proven highly successful.

How Venting and Housing Pressurization Affect Bearing Protector Seals

If in nonvented bearing housings the pressure starts to increase with an increasing temperature, then lip seals will release the excess pressure by lifting off. To release pressure, seal orientation is important. Lip seal frictional drag (and wear) would increase as housing pressure increases in Figures 8.1 and 8.4. In contrast, the microgap in the dual dynamic O-ring protector seals in Figures 8.1, 8.2(c), 8.4, and 8.6 would act as pressure relief passages.

Pressurizing a bearing cavity equipped with a dual-face magnetic seal (Figure 8.5)

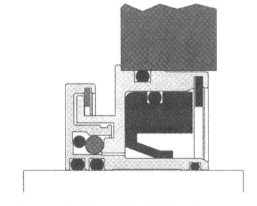

Figure 8.4: Advanced hybrid bearing protector seal.

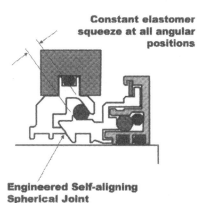

Constant elastomer squeeze at all angular positions

Engineered Self-aligning Spherical Joint

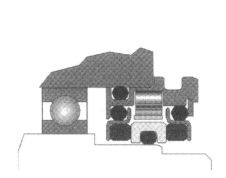

Figure 8.5: Dual-face magnetic bearing protector seal.

Figure 8.6: Self-aligning and extended axial movement bearing protector seal.

will add to the closing force of the rod magnets facing the left side of the rotor. The left Z-ring is close to the bearing, and the Z-ring to the right of the rotor is close to the ambient environment.

Housing-internal pressures up to 2 psig (13.8 kPa) over ambient are allowed in dual-face magnetic seals, with the seal manufacturer's concurrence. Pressure increase in nonvented bearing housings can be calculated using the combined gas law equation. This law says the ratio between the (absolute) pressure-volume product and the (absolute) temperature of a system remains constant. The pressure increase resulting from a known or assumed temperature rise in the same volume is simply $P_2 = P_1 T_2 / T_1$.

Because operating temperatures are rarely more than 100°F (~56°C) above standstill temperatures, the resulting pressure increase would stay below 2 psi. This Δp is too small to justify installing expansion chambers on most process pumps. (See Chapter 6 for comments on desiccant breathers and expansion chambers.)

COST JUSTIFICATION OVERVIEW

Lip seals will seal only while the elastomer material (the lip) makes full sliding contact with the shaft. Operating at typical shaft speeds on process pumps, lip seals show leakage after about 2,000 operating hours. To prevent contaminant intrusion, one would have to replace lip seals before they fail—twice a year for a certainty. In sharp contrast, modern rotating labyrinth seals incorporating the features shown in the lower right portion of Figure 8.1 would have a life typically exceeding 4 years (Ref. 5). Let's look at two cost comparisons:

Scenario 1—Replace Lip Seals before They Fail

To avoid shaft fretting, moisture intrusion, and premature bearing failure (assuming labor and materials to remedy a bearing failure costs $6,000), one would have to replace a $20 lip seal before it fails, say, twice a year. Labor ($500 per event) and materials (2 per pump) = $1,080 per year.

Alternatively, assume (just for the case of illustration) we replace two $200 modern dynamic O-ring rotating labyrinth seals after 4 years of operation. Labor: $125/yr, material: $100/yr; total: = $225 per year.

Benefit to cost ratio: 1,080/225 = 4.8

Scenario 2—Run Lip Seals to Failure

Lip seals are allowed to degrade and bearings fail after 2 years. Assume no production outage time, but the repair costs the plant $6,000, or $3,000 per year.

Again and as an alternative, assume (for the case of simple, ultraconservative illustration) we replace two $200 modern dynamic O-ring rotating labyrinth seals after 4 years of operation. Labor is $125/yr, material is $100/yr; the total is $225/yr. The benefit-to-cost ratio is 3,000/225, or 13.3.

Total cost of ownership and not just the as-purchased cost of the seal is important. More detailed cost justifications always include maintenance expenditures for oil changes. Note that labor, material, waste oil disposal, and other costs would have to be included. This would shift the picture even more in favor of modern bearing protector seals.

Advanced Bearing Housing
(Bearing Protector Seal) Summary

Lip seals have their place in disposable appliances and in machines that, for unspecified reasons, must frequently be dismantled. Lip seals do not measure up to the expectation of most intermediate and heavy-duty duty pump users.

Advanced rotating labyrinth seals are much preferred for reliable long-term operation of process pumps. In the span of several years between their market introduction and release of this book, none of the well-engineered models shown in Figure 8.1 had failed in operation and tens of thousands were in operation at release time.

A rigorous Weibull-WeiBayes statistical analysis demonstrated exceedingly long life for this field-repairable bearing protector seal. There is little (if any) risk of rotor-stator contact because the mass-symmetrical rotor is clamped to the shaft with two O-rings. The two-ring shaft clamping method imparts excellent stability and resists "rotor wobble."

Of course, bearing protector seals serve no purpose

- If oil contamination originates with oil ring inadequacies
- If used with unbalanced oilers (Chapter 6)
- If the oil is not kept at the proper oil level
- If the design causes or increases the risk of "black oil" formation (Refs. 2 and 3)

However, soundly engineered bearing protector seals excel by incorporating three attributes:

1. The assembly uses two O-rings (and not just one) to clamp the rotor to the shaft.
2. The rotor mass center is located centrally between these two O-rings and the rotor mass is not overhanging asymmetrically or outside of the two O-rings.
3. To qualify as soundly engineered bearing protector products and assuming they incorporate dynamic O-rings as sealing elements, these dynamic O-rings should act on a generously dimensioned, contoured surface and not on a sharp edge or groove.

WHAT WE HAVE LEARNED

Bearing housing protector seals can greatly improve both operating life and reliability of process pumps by safeguarding the cleanliness of the lubricating oil. As a rule, advanced products are estimated to extend pump bearing life four-fold.

Yet, not all bearing housing protector seals (bearing isolators) perform equally well. The user-purchaser must become familiar with the working principles of different offers and steer clear of vulnerable old-style configurations.

Thoroughly well-engineered products will not allow O-rings to contact sharp-edged components. They are superbly designed, with no chance of O-rings contacting sharp-edged grooves or other components. Although these designs are field-repairable and could be rebuilt by simply replacing a set of O-rings, thousands of these bearing housing protector seals have been in flawless service for many years.

References

1. Brink, Robert V.; Czernik, Daniel E., and Horve, Leslie A.; *Handbook of Fluid Sealing* (1993), McGraw-Hill, New York. (ISBN 10: 0070078270).

2. Bloch, Heinz P., and Alan R. Budris; *Pump User's Handbook: Life Extension*, 3rd Edition (2006), The Fairmont Press, Lilburn, GA. (ISBN 0-88173-629-9).

3. Bloch, Heinz P.; "Fighting black oil: how alternatives to oil rings may help prevent degradation"; *Turbomachinery International*, Jan./Feb. 2009 (Part 1), also March/April 2009 (Part 2).

4. Bloch, Heinz P., and Abdus Shamim; *Oil Mist Lubrication—Practical Application* (1998), The Fairmont Press, Lilburn, GA. (ISBN 088173-256-7).

5. Bloch, Heinz P. and Chris Rehmann; "Well worth the cost—bearing housing seals and value of reducing oil contamination," *Uptime Magazine*, Aug./Sept. 2008.

Chapter 9

Mechanical Sealing Options for Long Life

Mechanical seals are components that keep the pumped fluid (or "pumpage") in the pump casing. These seals are installed at the location where the shaft enters or leaves the casing, and their purpose is to prevent leakage. There are hundreds of styles, sizes, and configurations of seals. All of them use the underlying principles of stationary and rotating face combinations of Figure 9.1, and are more clearly depicted in Figure 9.13. Some are simple and inexpensive; others are complex and deserve special attention.

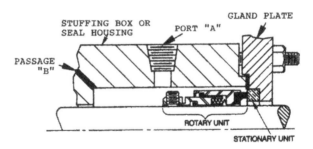

Figure 9.1: Operating principle and essential nomenclature of mechanical seals.

Conventional mechanical seals often apply spring pressure or some other closing force to the face of the rotary unit in Figure 9.1. However, many mechanical seals are designed with a single coil spring, several small coil springs, or pleated bellows applying a closing force to the stationary seal part. They are then called *stationary* mechanical seals. Application of spring-loaded closing force to the stationary seal part is advantageous when there is known shaft deflection and in shaft systems that operate at high peripheral velocity. Also, designs that place the springs away from the process fluid are generally preferred over designs that allow process fluid to contact the springs. All seal assemblies need some kind of vent provision, be it a port "A" or a simple vent passage "B" (Figure 9.1).

STILL USING PACKING?

Since about 1950, mechanical seals have almost totally replaced the more maintenance-intensive compression packing previously used in process pumps (Figure 9.2). But suppose you are presently in the process of converting to mechanical seals. In that case and until conversion work is completed, use the following installation and servicing sequence:

—Distinguish between different packing styles and materials; be sure
  to use the right one for the application.
—Preferably use a mandrel to wrap and cut new packing properly.
—Use the right tool to remove old packing.
—Remove corrosion products from shaft or sleeve surface.
—Coat ring surfaces with an approved lubricant.
—Insert one new ring at a time; then seat each ring using a proper tool.
—Install the packing gland—initially only finger-tight.
—Operate for run-in, initially allowing an uninterrupted stream of
  sealing water leakage.

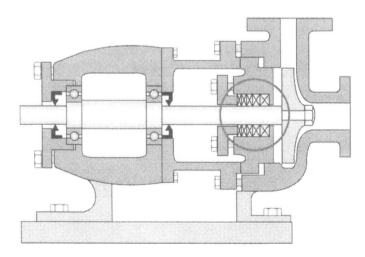

Figure 9.2: Compression packing used in a simple centrifugal pump.

—Tighten sequentially to allow a leakage flow of 40-60 drops per minute (inadequate leakage flow generates much heat and burns up packing).

GENERAL OVERVIEW OF MECHANICAL SEALS

Working with a competent seal manufacturer avoids mistakes and adds value. Reliability-focused users never view mechanical seals as a cheap standard commodity. These users select only experienced seal manufacturers and cultivate relationships based on mutual trust. Long-term customer service and consistent application of thorough engineering skills benefit both parties and are of much greater importance than short-term returns obtained from initially paying a low price. A single failure incident often causes price-related gains to vanish in a flash—both literally and figuratively.

Seal flush piping arrangements are often needed to create the most appropriate seal environment. The ones shown in this chapter and many other variants generally belong to two main groups:

1. "Flush"—Clean, cool liquid is injected into the seal chamber ("F1," Figure 9.16) to improve the operating environment. A second, smaller port is provided and normally oriented downward. It often serves as a leakage observation or vent opening ("Q/D," Figure 9.16).

2. "Barrier or Buffer" used with dual seals—A secondary fluid is fed to the space between two sets of mechanical seals to prevent ambient air from contacting the pumped fluid, to improve seal cooling, or to enhance safety.

It is worth mentioning that not all mechanical seals require special external flush arrangements. Still, for long and trouble-free life, the mating faces shown in Figure 9.1 must somehow be separated or cooled. The small gap between faces must be taken up by either liquid or gas. Different flush plans (flush schematics) accomplish that; they vary because they must accommodate different fluid parameters, conditions, and properties. All flush plans are described in vendor literature and industry specifications such as API-682 and ISO 21049. A few of them are shown in Figures 9.3 through 9.8; the ones we have selected here use American Petroleum Institute (API) plan numbers and are representative of the many different piping or flush plans available.

Note that the flush piping in most plans enters or exits at the top of the seal chamber, which then also allows vapors to be vented. In the few instances in which no such piped venting possibility exists, the pump user may have to drill a 0.125-inch (3-mm) diameter passageway (labeled "B" in Figure 9.1) at an angle of 15 to 45 degrees from the top corner of the seal chamber into the fluid space directly behind the impeller.

In preexisting pumps, certain upgrades or conversion to another (more modern) flush plan configuration are often advantageous. Where energy conservation issues and operating cost savings are deemed important, the newer seal configurations must be considered.

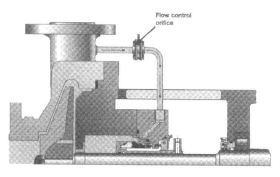

Figure 9.3: Recirculating (injecting) a product side stream (API Plan 11).

Recirculation of a product side stream (Plan 11, Figure 9.3) is common. The side stream should have a pressure of approximately 25 psi (173 kPa) higher than the pressure directly behind the pump impeller.

Axial holes have been drilled in the impeller discs of Figures 9.3 through 9.8. The pressure acting behind each particular impeller is thus kept near suction pressure. Drilling several axial holes also will influence the axial thrust produced by an impeller and may affect the load on the thrust bearing set. Recall that pressure (psi or N/mm^2) = force/area; hence, force = pressure multiplied by area. For optimum bearing life, this force cannot be too light (skidding risk was explained in Chapter 5) or too large (resulting in insufficient oil film strength and thickness).

A heat exchanger (Figure 9.4) will add to the cost of an installation, but it may be required in some services. Modern mechanical seals often include an internal pumping device—an important option that will be explained later in this chapter.

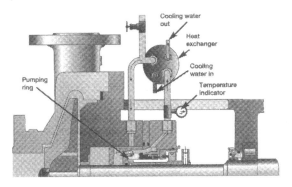

Figure 9.4: Product recirculation from seal chamber to heat exchanger and back to seal chamber per API Plan 23.

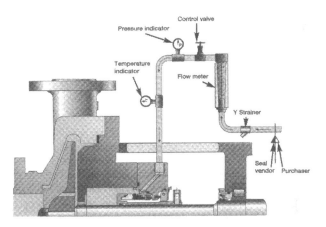

Figure 9.5: Injection of cool or clean liquid from external source into the seal chamber per API Plan 32.

Some applications favor the flush plan shown in Figure 9.5, although again, its ultimate suitability for a given application or service must be assessed. The liquid injected into the seal cavity will migrate through the throat bushing and into the pumpage. This dilutes the process fluid and the injected fluid will later have to be removed by evaporation. Because evaporative processes require considerable heat, the overall energy efficiency of the plant is inadvertently reduced simply because of a particular seal plan selection.

Dual seals use a barrier or buffer fluid to create a desirable seal environment (Figure 9.6). The cost of the seal auxiliaries shown here is clearly a

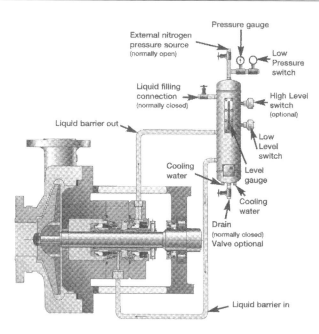

Figure 9.6: API Plan 53A—Pressurized barrier fluid circulation in outboard seal of a dual seal configuration. A pumping ring maintains circulation while running; thermosiphon action is in effect at standstill.

factor, and each application merits its own review. A connection to an external nitrogen source requires safeguards against the inadvertent backflow of barrier fluid into the external source of nitrogen. The possibility of nitrogen dissolving in the barrier fluid must be considered as well. In other words, seal selection requires study and experience.

Both Figures 9.7 and 9.8 are variations on our "go with advanced technology" theme and highlight why it is so important to have good cooperation between seal manufacturer and seal user-operator. Figure 9.8 (API Plan 62) is schematically shown again in Figure 9.9; many of these seals and flush plans have been converted to the far more effective water-management approach shown in Figure 9.10.

ALL FLUSH PLANS HAVE ADVANTAGES AND DISADVANTAGES

The many properties of fluids contacting the seal faces govern seal selection, but other considerations should be weighed as well. Long-term

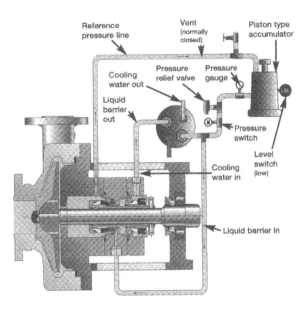

Figure 9.7: API Plan 53C—pressurized and cooled barrier fluid circulation in outboard seal of a dual seal configuration. A pumping ring (Figure 9-12) maintains circulation while running. The pressure is maintained and fluctuations are compensated in the seal circuit by a piston-type accumulator, upper right.

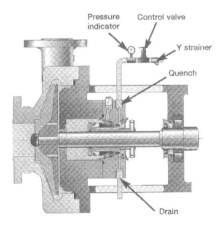

Figure 9.8: API Plan 62—An external fluid stream is brought to the atmospheric side of the seal faces using quench and drain connections (Q/D) in Figure 9.16. (This is an inefficient use of water.)

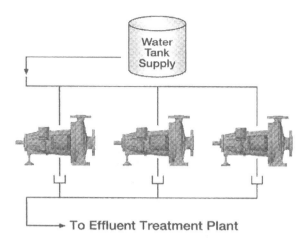

Figure 9.9: An energy-inefficient API Plan 62 in use at an older sewage treatment facility.

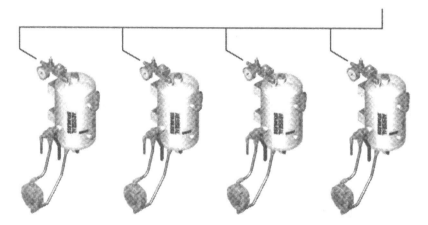

Figure 9.10: Four water management systems connect to the seal glands of four separate pumps in this sample illustration. There is no drain and no water is wasted.

reliability and savings in utilities should be given high priority. Conversely, low initial seal cost is rarely (if ever) a good indicator of the true value of mechanical seals and seal support systems.

Historically, Plan 23 has not received wide acceptance because of the obvious complications of applying old-style seal circulating devices or pumping rings. Use of modern computer-controlled manufacturing methods has

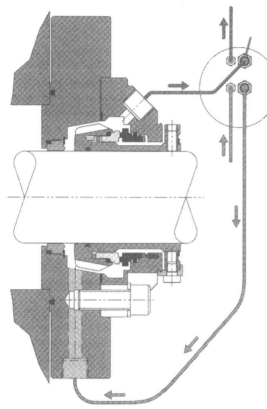

Figure 9.11: Plan 23 Cross section of a seal cavity with a bidirectional (tapered) pumping device. The flush product is pumpage that passed through the throat bushing on the left. After being cooled in a heat exchanger (upper right), this product reenters the seal region in the vicinity of the seal faces. A bidirectional tapered pumping device [See Figure 9.12] promotes relatively high flow rates and thus cooler face temperatures.

helped in implementing superior sealing technology.

Compact Plan 23 cartridge seals (Figure 9.11) are easily applied to both new and old pumps. These seals incorporate wide-clearance, bidirectional, and tapered pumping devices that are far less likely to make contact with seal-internal stationary parts than older, close-clearance pumping ring configurations.

ALWAYS OBTAIN THE FULL PICTURE

Seals are part of a pumping system and systems must be properly reviewed. This is demonstrated with API Plan 23 (Figure 9.11). The reviewer must first recognize that the seal assembly incorporates a throat bushing that will almost (but not fully) isolate the seal chamber from the pump's case. The small volume of liquid in the seal chamber is circulated through a local cooler. API Plan 23 is used on hot applications and minimizes the load on the heat exchanger. It needs to cool only the heat generated by the seal faces and the heat that has migrated through the seal chamber casing. An effective pumping or circulating device (Figure 9.12) is at the heart of Plan 23. (See also Figures 9.4, 9.16, and 9.17).

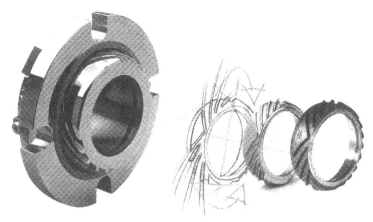

Figure 9.12: A bidirectional tapered pumping ring assembled (left) and shown separate (right).

SEAL CHAMBER PUMPING RING (CIRCULATING DEVICE) TECHNOLOGIES

Many pumping rings found in mechanical seals are based on a straight vane or paddle-type configuration (Figure 9.13). Typical pump-around flow rates achieved with traditional pumping rings are low. They will function only in the plane in which the ports and the straight vanes (paddles) are located. Tangential porting will be required, and in many instances, little or no liquid is induced to flow continually over the seal faces.

Pumping screws (Figure 9.14) are considerably more efficient; however, they must rely on a dimensionally close clearance gap between screw periphery and housing bore. This close gap can be a serious liability in situations in which shaft deflection or concentricity issues exist. Designs with a close gap will not comply with the API 682 (2002) requirement of a minimum radial clearance of 0.060 inches (1.5 mm). But screw devices with large clearance gaps (Figure 9.14) have poor efficiency and can be as ineffective as the straight-vane pumping rings of Figure 9.13.

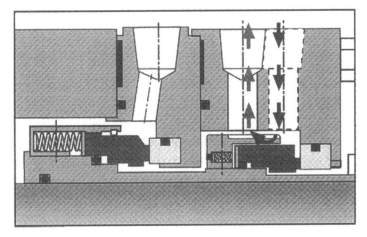

Figure 9.13: A relatively ineffective straight-vane pumping ring.

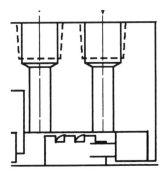

Figure 9.14: Pumping screw found in some dual seals. To be effective, these auger-type helical screws will require close clearances, but close clearances introduce risk factors.

Traditional pumping screws (Figure 9.14) are, of course, unidirectional. This inherent unidirectionality leaves ample opportunity for human error on between-bearing pumps; recall that left-hand devices are required on one end and right-hand devices are required on the other end of the shaft.

Close radial clearances between counter-rotating surfaces can lead to component contact and galling. If a stainless steel rotary component contacts a stainless steel stationary component, then galling will occur.

Some dual seals applied in industry are designed with radial clearances in the order of 0.010 and 0.020 inches (0.25 mm and 0.5 mm). This contradicts best-practice guidelines and technical logic because pumps operating away from best efficiency point (BEP) and most single volute pumps are known to undergo a measure of shaft deflection.

Whenever the dual seal radial clearance is less than the throat bushing clearance in the pump, a close-clearance helical screw device will be the first to make contact. All these potential issues are avoided with open-clearance, bidirectional, and tapered pumping rings.

Lessons Apply to Many Services

We chose here to summarize the seal topic by using a few schematic representations. In API Plan 21 (Figure 9.15), process fluid is diverted from the discharge of the pump, sent through a restriction orifice as well as a seal flush cooler, and then routed into the seal chamber. The seal flush cooler is removing heat from the process stream. If the production process demands a hot fluid, then such heat removal will benefit only the seal. In that case, cooling the flush stream would reduce the overall process energy efficiency.

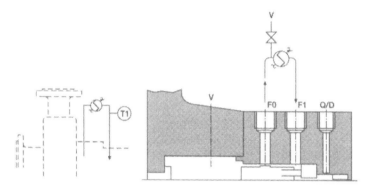

Fig 9.15: Recirculation from pump discharge through restriction orifice and cooler to seal chamber (API Plan 21).

Plan 23 can offer benefits of improved vapor pressure margin in seal chambers, thereby extending seal reliability. The reduced working temperature of Plan 23 operation has prevented coking on the atmospheric side of many mechanical seals in hydrocarbon services. This, of course, enhances seal life. There are only a few applications in which Plan 21 is still preferred over Plan 23 and a competent seal manufacturer can point them out.

Self-contained water-management systems (Figure 9.17, also Figure 9.10) are easily cost-justified in applications that must conserve water. Self-checking hydraulic sensing valves are incorporated in these systems.

In essence, retrofits of water management systems are conversions from Plan 32 (Figure 9.18) to Plan 53A (Figure 9.19). These conversions often make considerable economic sense and have even improved plant output in distilleries and paper plants.

Upgrading to Plan 53A (Figure 9.18) was implemented on evaporator pumps at distillery units. The small pressurized tanks in Figure 9.16 con-

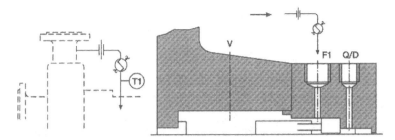

Figure 9.16: API Plan 23-Recirculation from a pumping device in the seal chamber through a cooler and back to the seal chamber.

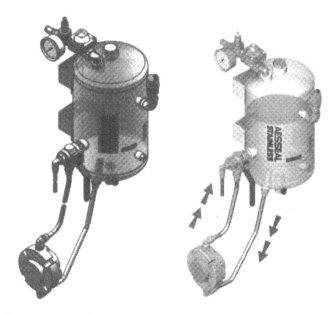

Figure 9.17: Self-contained water-management systems are replacing many highly inefficient Plan 62 systems that were formerly used.

tained 10 liters of barrier fluid and were installed at each pump. In one facility, a three-pump conversion allowed syrup production to be increased from 88 to 98 tons/hr, and the plant was subsequently able to operate fewer hours while still meeting full capacity demand.

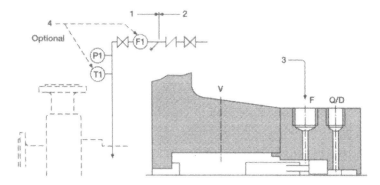

Figure 9.18: API Flush Plan 32-Flush is injected into the seal chamber from an external source.

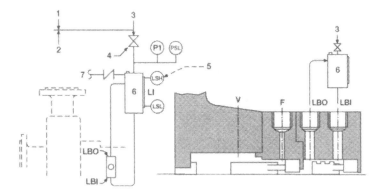

Figure 9.19: API Flush Plan 53A-A pressurized external barrier fluid reservoir supplies clean fluid to the seal chamber at a pressure greater than that of the process (pumpage). An internal pumping device provides circulation.

WHAT WE HAVE LEARNED

• Phasing out any remaining packed stuffing boxes and upgrading to well-designed mechanical sealing alternatives makes economic sense. Follow good stuffing box packing practices until conversion to mechanical seals is complete.

• It is necessary to ascertain that vapors or gases (air) will be vented from the seal cavity. A small hole may have to be drilled in the pump cover or seal housing to accomplish this venting when "dead-ended" seal arrangements are used.

- Plan 23 is generally preferred for hot water services, particularly boiler feed water. This plan is also desirable in many hydrocarbon services in which it is necessary to cool the fluid to establish the required margin between fluid vapor pressure and seal chamber pressure.

- In just a few special cases, Plan 21 is preferred over Plan 23. Work with a competent seal manufacturer to identify these exceptions to the rule.

- Plan 53A (self-contained water-management systems) are often vastly superior to inefficient Plan 32 configurations.

- Maintaining an adequate vapor pressure margin helps protect the seal faces against localized boiling of the process fluid at the seal faces. This can cause loss of seal-face lubrication and subsequent seal failure.

- Lowering the flush fluid temperature is always preferable to pressurizing the seal chamber.

- Bidirectional pumping devices are far less likely to make contact with internal seal components under conditions of pump stress. Use them whenever possible.

- Competent mechanical seal manufacturers are a valuable reliability improvement source. Make good use of their expertise and engineering know-how.

Bibliography

1. API-682 (ISO 21049), American Petroleum Institute, Washington, DC, "Pumps-Shaft Sealing Systems for Centrifugal and Rotary Pumps," 3rd Edition, International Standard, Sept. 2004

2. Francis, D.W., M.T. Towers, and T.C. Browne; "Energy Cost Reduction in the Pulp and Paper Industry—An Energy Benchmarking Perspective," Pulp and Paper Research Institute of Canada; the Office of Energy Efficiency of Natural Resources, Canada, 2002.

3. AESSEAL plc. Rotherham, UK, and Rockford, TN; API Flush Plans (marketing literature/laminated booklet).

4. Gurrfa, Angel (Secretary-General, OECD); "Water: How to manage a vital resource," OECD Forum; May 14-15, 2007, Paris, www.oecd.org.

5. BNXS01: Carbon Emission Factors for UK Energy Use; Defra Market Transformation Program, Version 2.2, July 10, 2007.

6. Smith, Richard; "Reconciling requirements in API-682 dual seal design configurations," *Hydrocarbon Processing*, February 2009.

Chapter 10

Pump Operation

Within the boundaries of a slim volume on pump wisdom, operating issues cannot be overlooked. Occasional reviews, comparisons, and periodic updates (or reaffirmations) of a facility's present centrifugal pump startup procedures are almost certain to reduce failure risk. Consider these review activities highly recommended.

Reliability-focused pump users view pump starting, performance monitoring, and shutting down as operator-driven reliability functions that use the following steps and sequences:

STARTING CENTRIFUGAL PUMPS

1. Arrange for an electrician and a machinist/specialist to be present when a pump is initially commissioned. Ascertain that large motors have been checked out.

2. Close or almost close the discharge valve and fully open the suction valve. Except for axial flow pumps (pumps with high specific speed), the almost-closed discharge valve (leave it about 10% open) creates a minimum load on the driver when the pump is started.

   Assuming that the motor inrush current allows and that the motor will not kick off, the discharge valve may be just "cracked open" (again, about 10% open) before the pump is started. (A fully closed gate valve can be difficult to open because high-pressure forces the sliding valve parts into the surrounding stationary parts).

3. Ascertain that seal flush is "lined up" and all related checks and procedures have been complied with.

4. Make sure the pump is primed. Opening all valves between the product source and the pump suction should get product to the suction, but it does not always ensure that the pump is primed.

Only after ascertaining that fluid emissions are not hazardous or are routed to a safe area, open the bleeder valve at the top of the pump casing until all vapors are exhausted and a steady stream of product flows from the bleeder. It may be necessary to open the bleeder again when the pump is started or even to shut down and again bleed off vapor if pump discharge pressure is erratic.

> **Note:** Priming of a cold service pump may have to be preceded by "chill-down." A cold service pump is one that handles a liquid that vaporizes at ambient temperatures when under operating pressures. Chilling down a pump is similar to priming in that a casing bleeder or vent valve is opened with the suction line open. The following additional factors should be considered for cold service pumps:

- Chilling a pump requires time for the pump casing to reach the temperature of the suction fluid.

- The chill-down vents are *always* tied into a closed system.

- On pumps with vents on the pump case and on the discharge line, open the vent on the discharge line first for the chill down and then open the pump casing vent to ensure that the pump is primed.

Should it be necessary to have a cold service pump chilled down and ready for a quick start (e.g., refrigerant transfer pumps during unit startup); in that case, the chill-down line can be left cracked open to get circulation of the suction fluid.

5.   If a minimum-flow bypass line is provided, then open the bypass. Be sure the minimum-flow bypass is also open on the spare pump if it starts automatically.

6.   Never operate a centrifugal pump without liquid in it and never operate it with both the suction and discharge valves closed.

7.   Check lube oil and seal pot level (assumes dual seals).

8.   Start the pump. Confirm that the pump is operating by observing the discharge pressure gauge. If the discharge pressure does not increase, then stop the pump immediately and determine the cause.

9. Open the discharge valve slowly while watching the pressure gauge. The discharge pressure will probably drop somewhat, and then level off and remain steady. If it does not drop at all, then there is probably a valve closed somewhere in the discharge line. In that case, open the discharge valve.

    Do not continue operation for more than 5 seconds with a closed discharge valve or a blocked line.

10. If the discharge pressure drops to zero or fluctuates widely, then the pump is not primed. Close the discharge valve and—if safe—again open the bleeder from the casing to exhaust vapor. If the pump does not pick up at once, as shown by a steady stream of product from the bleeder and steady discharge pressure, then shut down the pump and check for closed valves in the suction line. A dry-running pump will rapidly destroy itself.

11. Carefully check the pump for abnormal noise, vibration (using vibration meter), or other unusual operating conditions. An electrician and machinery engineer should be present when pumps are started up for the first time (i.e., upon being initially commissioned).

12. Be careful not to allow the bearings to overheat. Recheck all lube oil levels.

13. Observe whether the pump seal or stuffing box is leaking.

14. Check the pump nozzle connections and piping for leaks.

15. When steady pumping has been established, close the start-up bypass and chill-down line (if provided) and check that block valves in minimum flow bypass line (if provided) are open.

16. Recheck lubricant quantity (level) and bearing housing temperatures. If it is too hot to the touch (the human hand does not tolerate exposure to more than 170°F or 76°C for over 5 seconds), ask for an exact measurement using infrared or surface pyrometer.

SURVEILLANCE OF PUMP OPERATION

1. Especially at the start of pumping, but also on periodic checks, note any abnormal noises and vibration. If excessive, then shut down.

2. Note any unusual drop or increase in discharge pressure. Some discharge pressure drops may be considered normal. When a line contains heavy, cold product, and the tank being pumped out contains a lighter or warmer stock, the discharge pressure will drop when the volume in the line has been displaced.

    Also, discharge pressure will drop slowly and steadily to a certain point as the tank level is lowered. Any other changes in discharge pressure while pumping should be investigated. If not explainable under good operating conditions, then shut down and investigate thoroughly. Do not start up again until the trouble has been found and remedied.

3. If lubricated by oil mist, then periodically check the oil mist bottom drain sight glass for coalesced oil or water and drain if necessary.

4. On open (old style) oil mist systems, regularly check both the oil level and the stray oil mist flow from vents or labyrinths.

5. Seal oil pots need to be checked regularly for the correct level. Refill with fresh sealing liquid (a special oil, propylene glycol, or methanol, as specifically required and approved for the particular application).

6. Periodically check for excessive packing leaks, mechanical seal leaks, or other abnormal losses. Also, check for overheating of packing or bearings. When in doubt, do not trust your hand. Use an infrared gun or a surface pyrometer instead. Note that excessive heat may cause bearing failure and may result in costly and hazardous fires.

7. At a minimum, switch on the second or "spare" pump at least once a week and run for 2 hours to prevent the bearings and mechanical seals from seizing.

    Realize, however, that "best practices plants" have determined monthly switching of the "A" and "B" pumps to be the most appropriate (optimum) and cost-effective, long-term operating mode.

8. When a pump has been repaired, place it in service as soon as possible to check its correct operation. Arrange for a machinist to be present when the pump is started up.

CENTRIFUGAL PUMP SHUTDOWN

Again, reliability-focused pump users consider it worthwhile to re-assess their shut-down procedures occasionally. These are among the many operator-driven reliability functions in which the three primary job functions (i.e., operations, maintenance, and project-technical) merge and in which there needs to be a consensus among these three job functions.

Reliability-focused pump users consider using the following procedures:

Special note: On hot service pumps and on shutting down, commission ("line up") the "restart liquid fill" piping. Make sure the mechanical seal region is not exposed to viscous pumpage.

1. Close the discharge valve. This takes the load off the motor and may also prevent reverse-flow through the pump.

2. Shut down the driver.

3. If the pump is to be removed for mechanical work, then close the suction valve and open the vent lines to flare or drain, as provided. Otherwise, leave the suction valve open to maintain the pump at the correct operating temperature.

4. Shut off steam tracing if any. Continue oil mist lubrication if provided.

5. Shut off cooling water (which is provided on sleeve bearings), sealing oil, and so on, if the pump is to be removed for mechanical work.

6. At times, an emergency shut-down may be called for. If you cannot reach the regular starter station (e.g. in case of fire), then stop the pump from the remote starter box, which is located some distance away and is usually accessible. If neither the starter station nor the remote starter box can be reached, then call the electricians.

    Still, do not—as a part of regular operations—stop pumps from the remote starter box. Use the regular starter station instead.

Please note that these procedures are of a general nature and may have to be modified for nonroutine services. Always review pertinent process data, as applicable. Rewrite these instructions in concise sentences if they are to become part of checklists that operators are asked to have on their person while on duty.

WHAT WE HAVE LEARNED

- Generalized pump start-up and shut-down procedures apply to the majority of radial impeller centrifugal process pumps.

- Specialized procedures are needed for mixed flow and axial flow pumps.

- Precooling and/or preheating impose special requirements on cold and hot service pumps.

- Venting requirements can differ from service to service, making it important to view pumps as part of a system.

- The operators are the eyes and ears of the plant. They are trained to be the "first responders" and must accept the responsibility to be the first to notice deviations from the norm.

- The operators' job must include data collection and equipment surveillance. Once they spot a deviation from normal pump behavior, they are empowered to ask other job functions within the company to assist in analyzing the root cause.

Chapter 11

Impeller Modifications and Pump Maintenance

Please recall that this text is not intended as a pump maintenance book. Many pump manufacturers have compiled recommendations on pump maintenance, and these vary from one manufacturer to the next. They generally differ among pump types and models. Yet, there are some common threads or essentials to be considered.

MAINTENANCE ESSENTIALS

Maintenance is to prevent deterioration and, if necessary, to restore things to the as-sold or as-installed state. Routine preventive maintenance of process pumps is generally limited to oil replacement and replenishing. Next in line is mechanical seal maintenance, including seal replacement.

Chapters 6 and 7 dealt with lubricant application, cooling, and lube types. Lubricant replacement (oil change) is done four times per year at a particular U.S. refinery—one that clings to the outdated and wasteful tradition of using cooling water on its pumps. Another refinery with soundly engineered constant level lubricators, advanced bearing housing protector seals and premium quality synthetic lubricants (ISO VG 32 and 68) changes oil every 3 years.

Of course, oil change issues do not exist at facilities that use pure oil mist instead of liquid oil. Some pump manufacturers decline to comment on the fact that no maintenance is needed for oil mist lubricated bearings. Declining to comment may simply be the result of not having updated one's knowledge base.

After a repair has been completed, proper installation and its verification are always needed. But additional maintenance tasks vary from plant to plant. These include condition monitoring by periodically comparing head versus flow data—an operator-technical function—and spotting deviations before they result in process-operational and safety issues.

It's always good to remember that a single deviation rarely causes pumps to fail. Regrettably, people assume that they can get away with another deviation, and a third one, and a fourth one. After a while, living with these deviations becomes the "new normal." Then, suddenly, just one more deviation occurs. Now the pump fails massively and puts human lives and physical assets at great risk. The hunt for culpable parties begins and the legal profession is mobilized. Interviews and depositions are being arranged. Orders of magnitude more money is spent than what it would have cost to avoid the problem in the first place. How, then could the problem have been avoided? By superior maintenance!

SUPERIOR MAINTENANCE REQUIRES UPGRADING

Suppose there were a design flaw or some other hidden, elusive issue that is responsible for causing repeat failures, short run lengths of a pump, or whatever else. More than traditional maintenance effort may be needed, which gets us to the subject of superior maintenance. Superior maintenance is a reliability-improvement task. The best-performing or best-of-class companies are not repair focused; they are reliability focused. These are companies that have given some of their personnel the task of determining whether upgrading to a better component is feasible.

If an upgrade is feasible, then the contributor is asked to calculate its cost. A best-performing company would then ask what this upgrade effort will be worth, perhaps in a simple benefit-to-cost or payback estimate.

A good example involving upgrade issues is shown in Figure 11.1. The "cooling fan" in the left illustration is simply too small to be of any value. The well-dimensioned fan shown on the right will probably be effective. Recall, however, that an earlier chapter highlighted the potential drawbacks of applying cooling only to bearing outer rings. The hot inner rings may thermally expand and cause bearing preload increases that will then reduce bearing life.

Suppose an effective fan was used. In that case, the entire design must be reviewed for pressure differences surrounding the bearings. These differences might affect lube oil flow or oil level conditions in the adjacent bearing set. Therefore, many bearing housings must be upgraded to reduce failure

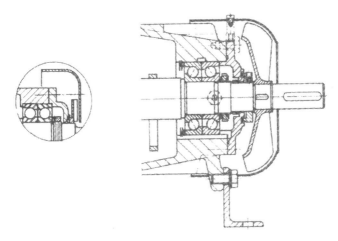

Figure 11.1: Ineffective "cooling fan" (left) and a reasonably effective fan (right).

risk. In view of our previous chapters and comments on the deletion of bearing cooling, the best course of action would be to discard both fans in Figure 11.1. Also, it would be prudent to do the following:

(a) Implement means of pressure equalization across the thrust bearing set
(b) Use an appropriate synthetic oil
(c) Consider the various shortcomings of using oil ring(s) shown on the right of Figure 11.1
(d) Question and then verify the axial load capacity of the snap ring in the housing bore shown on the right side of Figure 11.1

IMPELLER UPGRADING WITH INDUCERS

Three of the many thousands of impeller configurations and geometries available are illustrated in Figure 11.2. Closed and open impellers were briefly mentioned in earlier chapters; an inducer-type impeller is shown in the center of the picture.

Inducers are often custom-designed for a particular pump application. They lower the net positive suction head required (NPSHr) of an impeller but do so only in the vicinity of best-efficiency flow (BEP). If pumps with inducer-equipped impellers are operated at flows substantially above or below BEP, then their NPSHr may actually increase compared with that of a standard impeller not so equipped. Instead of the continually increasing slope of the NPSHr curve shown earlier in Figure 1.4, the NPSHr curve associated with an inducer would probably have a distinct U-shape.

Figure 11.2: Closed impeller (left), impeller with inducer (center), open impeller (right).

DISTANCE FROM IMPELLER TIP TO STATIONARY INTERNAL CASING COMPONENTS

Any particular process pump is designed with a casing that, of course, surrounds the pump rotor. For this rotor and its impeller to turn freely while, at the same time, accommodating a certain amount of shaft run-out and shaft deflection (see earlier Figure 1.9), there needs to be a rather liberal clearance.

In practice, one makes a distinction between two different gaps shown in Figure 11.3: Gap "A" and Gap "B." Gap "A" is the distance from the impeller disc and cover (the cover is sometimes called a shroud) peripheries to the nearest casing or other stationary part. That part is usually the tongue or cutwater. Some pumps are designed with casing internals called diffuser-style or vaned; others are simply an unimpeded stationary passageway for the fluid leaving the impeller. Regardless of casing-internal construction, a gap of 50 mils (0.050 inches or 1.2 mm) is chosen for best possible efficiency and to minimize leakage-induced flow turbulence.

Gap "B" is the average distance from the impeller vane tips (the diameter D') to the casing-internal stationary part in which the fluid coming off the impeller vane tips enters a stationary passageway. This part of the stationary passageway is usually called the cutwater, or volute tongue. Recommended (radial) minimum, maximum, and preferred values for Gap "B" are

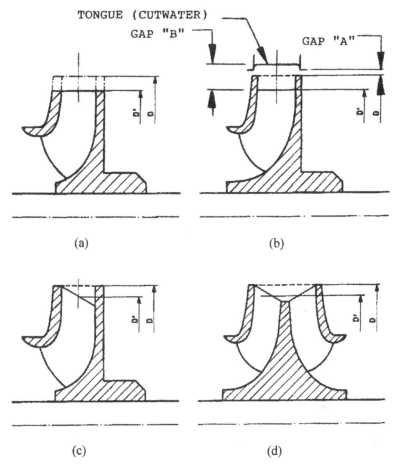

Figure 11.3: Gap designations and impeller trim methods, including vane tips only (left); across entire outside diameter (center); oblique cut (right), based on Ref. 1.

given, in Table 11.1, as a percentage of the impeller radius. In large pumps (typically pumps with drivers in the over 250-kW range), a gap that is too small will often increase vibration amplitudes, whereas a gap that is too large will result in a loss of efficiency.

As an alternative to Gap "B" modifications, volute chipping—the removal of a small amount of metal from the tongue or cutwater—is sometimes feasible (Ref. 1).

Table 11.1: Recommended radial gaps for process pumps

Type	Gap "A"	Gap "B" Percentage of Impeller Radius		
		Minimum	Preferred	Maximum
Diffuser	50 mils	4%	6%	12%
Volute	50 mils	6%	10%	12%

Impeller Trimming

Virtually every textbook on pump engineering contains guidelines on impeller diameter reductions ("trimming") needed to affect the rate of flow permanently and the head created by a constant-speed centrifugal pump. Yet, many of these texts lead to the erroneous assumption that suitable diameter reductions are made by simply machining uniformly across the full periphery—impeller cover (shroud), vane tips, and disc. These texts use Euler's fan law equation, a mathematical relationship in which n, D, Q, H and P are, respectively, rotor (impeller) rpm, impeller diameter, flow, head, and power demand. In the following equation, the subscript 1 refers to the original value and subscript 2 indicates the new value (Ref. 2):

$$\frac{Q_1}{Q_2} = \frac{n_1 D_1}{n_2 D_2} \qquad \frac{H_1}{H_2} = \frac{n_1^2 D_1^2}{n_2^2 D_2^2} \qquad \frac{P_1}{P_2} = \frac{n_1^3 D_1^3}{n_2^3 D_2^3}$$

However, the various assumptions on which these relationships are based are rarely giving precise answers. Flow angles and the resulting velocity relationships are being disturbed by trimming. Experience shows that in real-world situations a reduction of impeller diameter greater than 15% of the original full diameter should not be allowed. To be on the safe side and not to cut too much, a prudent pump specialist will look at the mathematically derived diameter reduction (sometimes called the "fan law") and then make a trim cut of only 70% of what the Euler-based fan law math requires.

Suppose, as an example and with no speed change, we wanted a pump head reduction from previously 680 feet to a new head of 580 feet. Suppose also that we had an original diameter D_1 of 13.00 inches and wished to determine the new diameter D_2. After doing the algebraic transposing and square root extracting, we quickly find a new diameter of 12.14 inches—93% of the original. Using the 70% rule, we would now remove not 13.00 − 12.14 = 0.86 inches but, instead, only (0.7)(0.86) = 0.6 inches. In other words, field experience tells us the impeller should be trimmed to 12.4 inches. It may not be the exact needed diameter, but we will have avoided the risk of trimming off too much metal.

Good practice would be to trim only the vane tips (Figure 11.3, left illustration) and to leave both the impeller cover (sometimes called impeller shroud) and the disc at the recommended Gap "A" diameters. Best practice would be to cut obliquely (see Figure 11.3, right side) for the greatest structural support and to ward off resonant vibration. Vibration of unsupported regions can lead to cracks and failure.

With oblique cuts, the average diameter D' is used as the relevant diameter D_2 in the Euler equation. Either oblique cutting or partial removal of unsupported regions (called "scalloping" in Figure 11.6) is done to reduce the risk of fatigue failure.

Impeller Wear Rings

Modern process pump impellers are commonly fitted with plain wear rings. Wear rings separate regions of high pressure (discharge) from regions of low pressure (suction). Wear rings are considered replaceable parts, and the gap between stationary (casing-mounted) and rotating (impeller-mounted) wear rings should follow the guidance found in API-610.

Most wear rings are plain cylindrical, although step and labyrinth types (Figure 11.4) are occasionally used in efforts to reduce leakage flow. Their effectiveness does not appreciably differ from that of a plain wear ring. Time and effort needed to gain marginal improvement by implementing clearances 1 and 2 ("b" on right of Figure 11.4) is rarely worth it.

Wear rings with enhanced contours (Figure 11.5, right side) make good use of the simple hydraulic vortex principle and are considered effective, low-cost means of optimizing impeller performance (Ref. 3).

Vane Tip Overfiling and Underfiling

All impeller tip cuts and especially trim cuts across the entire impeller diameter (as done in the center illustration in Figure 11.3) produce

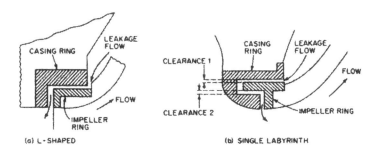

Figure 11.4: L-shaped ("a," on left) and labyrinth-shaped impeller wear rings ("b," on right) Source: Ref. 2.

Figure 11.5: Conventional wear ring profiles (left) and enhanced contour wear rings (right). Source: Ref. 3.

blunt vane tips and hydraulic disturbances. These disturbances can be lessened by modifying the blunt vane tips by overfiling or underfiling (Figure 11.6).

Removing metal from the leading edge of a vane tip is called "overfiling." The region from which metal is to be removed by tapering and blending-in is shown as "L." A guideline value for length "L" is 30% of the impeller radius R (or 15% of the impeller diameter). The vane tip width or thickness would be reduced to roughly 50% of its previous blunt edge width and metal removed to blend in without creating a step or ridge. To not compromise strength and resistance to erosion, even a "sharp" tip should not be thinner than 8 mm (~5/16 inch).

"Underfiling" is the term used for metal removal from the trailing edge of the impeller vanes. About 4% additional head can thus be gained near the best efficiency point (BEP), and the H-Q curve is shifted slightly to the right. In other words, the liquid channel has been marginally enlarged and impeller performance is enhanced somewhat.

We chose a semiopen impeller (Figure 11.6) for ease of illustration only. It should be noted that scalloping—the removal of unsupported material between vanes—is generally done on the shroud (cover) of a closed impeller. This scalloping reduces resonant vibration and fatigue cracking risks (Ref. 1).

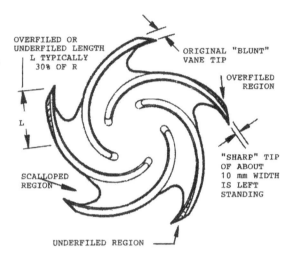

Figure 11.6: Vane modifications suppress flow disturbances and vibrations.

WHAT WE HAVE LEARNED

Doing things right would include allowing no shortcuts on impeller technology.

- Observing gaps "A" and "B" and doing oblique impeller trims is best practice. Inculcating a mindset that is uncompromising will pay back handsomely for decades. Best-of-class performers easily reach pump mean-times-between failure (MTBFs) of 6, 8, and even 10 years.
- Scalloping reduces the risk of crack formation and fatigue failure.
- Enhanced contour wear rings are simple and relatively effective performance enhancers.
- Some cooling fans are far too small to be of any value, and cooling fans are often unnecessary on pumps.
- Making it a practice to allow deviations and accepting them will soon make deviations the "new norm." Allowing deviations to add up is a

huge risk; it will certainly prevent a plant from ever becoming a best-of-class performer.

References

1. Dufour, John W., and William Ed Nelson; *Centrifugal Pump Sourcebook* (1993), McGraw-Hill, New York, NY. (ISBN 0-07-018033-4).
2. Karassik, Igor J., William C. Krutzsch, Warren H. Fraser, and Joseph P. Messina; *Pump Handbook*, 2nd Edition (1985), McGraw-Hill, New York, NY. (ISBN 0-07-033302-5).
3. Kurtz, Reinhard, and Katrine Abraham; "Special systems for coating cars," *World Pumps*, May 2010.

Chapter 12

Lubrication Management

A listing of the four or five worst enemies of proper and reliable lubrication must include lubricant contamination, viscosity, and application deficiencies. These three parameters are interrelated. For instance, water entering as an outright contaminant or in the form of air saturated with moisture often causes additive depletion. The lubricant then rapidly degrades and bearings fail prematurely. Water can also act as the catalyst for sludge formation and often has an undesired effect on viscosity. Oil rings tend to malfunction in oils of inappropriate viscosity.

Profitability-minded plants pay attention to these facts and thoughtfully manage all of their lubrication practices, including oil storage and oil transfer. With thoughtful lube management, they reap substantial benefits from the enhanced pump reliability that results.

HOW BAD IS WATER CONTAMINATION?

Even relatively small amounts of water contamination tend to affect bearing performance. The life of a rolling element bearing with as little as 100 ppm of free water in the lubricant (1 teaspoon of water in 13 gallons of oil) will be only 40% of the life of the same bearing with 25 ppm free water present (Ref. 1). Also, most major bearing manufacturers have published estimates on the effects of lubricant contamination (Ref. 2).

Considerable time is often wasted in arguments over how much water should be tolerated in pump bearing housings. It would be far more productive to locate the source of water intrusion and to cut off the source. Certainly, the amount of water drained from the pump bearing housing in Figure 12.1 is totally unacceptable.

In all instances, eliminating unsuitable transfer containers (Figure 12.2) must be made one of the first orders of business.

Figure 12.1: Excessive amount of water being drained from a pump bearing housing.

Figure 12.2: Unsuitable lube transfer containers.

Galvanized oil transfer containers are frequently attacked by certain lube oil additives. Good lube management practices mandate the use of approved dispensing containers. Of course, even the best purpose-designed plastic containers can be defeated by careless handling. This is illustrated in Figure 12.3.

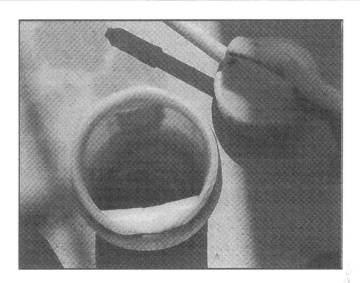

Figure 12.3: Careless handling of an approved transfer container can still cause extreme particulate contamination (Ref. 3).

Lube oil contamination can be kept in check in several ways. Modern bearing protector seals will prevent the ingress of water, steam, and other airborne contaminants (see Chapter 8). Either dual face (contacting) magnetic seals or advanced rotating labyrinth style (non-contacting) bearing protector seals are available. Both types make economic sense and merit consideration in new and existing (retrofit-to-upgrade) installations.

AVOID SOLIDS CONTAMINATION

In Figure 12.4 and similar illustrations published in technical texts, major multinational bearing manufacturers alert us to the life reduction risk associated with lube oil contamination. The ratio of operating viscosity of the lubricant (see Figure 7.2) over its rated lubricant viscosity (see Figure 7.1) is calculated as v/v_1 and entered on the x-axis.

Three relative stages of oil cleanliness are indicated by three different regions. Region I is applicable only to situations combining utmost cleanliness in the lubrication gap with moderate bearing loading. This situation is rather unrealistic for process pumps. Region II is where a high degree of cleanliness is maintained; this condition is assumed achievable with modern

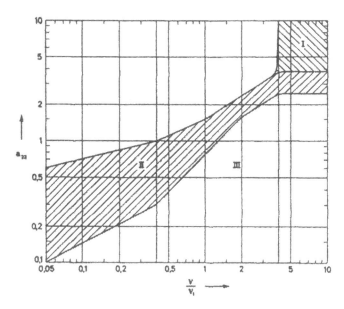

Figure 12.4: Contamination adjustment factor "a_{23}" versus viscosity ratio v/v_1, (per Ref. 2)

bearing protector seals (see Chapter 8). Finally, assume Region III might represent bearing housing interiors exposed to an industrial environment or bearing housings *without* effective protector seals. A resulting life adjustment multiplier or factor "a_{23}" is displayed on the y-axis.

Using Figure 12.4 and for, say, $v/v_1 = 0.5$, one would obtain $a_{23} = 0.3$ in Region III, and $a_{23} = 1$ in Region II. In essence, using effective bearing protector seals (see Chapter 8) yields an anticipated three-fold bearing life improvement.

AVOID QUESTIONABLE OIL STORAGE AND TRANSFER PRACTICES

Questionable drum storage practices are shown in Figures 12.5 and 12.6. Changes in ambient temperature cause rainwater to be drawn into the drum by capillary action. Capillary action takes place even in the case of seemingly tight bungs (Figure 12.5) as long as rainwater is allowed to collect in the drum top. The drum top will bulge outward (convex shape) when the

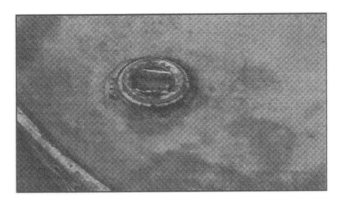

Figure 12.5: Rainwater will enter through this closed bung via capillary action.

surrounding air warms up; it will bulge down (concave shape) when the air cools and a slight vacuum is produced in the vapor space inside the drum. The resulting suction effect draws in water.

Open and up-turned vents in Figure 12.6 invite entry of both moisture and particulate contamination. Each drum contains $1,200 worth of synthetic lubricant that is now rendered unserviceable. One solution is shown in Figure 12.7 where outdoor drum storage has been replaced by

Figure 12.6: Unacceptable outdoor lube storage where up-turned open vents allow contaminants and moisture to enter.

indoor minibulk storage in sheltered or similarly protected locations. Bulk tanks and drums are best placed under cover.

Contamination control of bulk storage containers makes it highly desirable to put oil drums indoors. In the event that outdoor storage is chosen, the storage area must be covered by a suitable roof and side panels to ward off rain and snow.

Largely self-contained lubrication work centers (Figure 12.8) are available for clean oil storage and oil dispensing. These also merit serious consideration (Ref. 3). They replace traditional storage in cumbersome and generally inefficient 55-gallon drums. Modern lubrication work centers incorporate an interchangeable series of frames, pumps, filters, and storage modules. There also are spill containment cans and other suitable keep-clean provisions.

The new work centers should be preferred over antiquated tank and drum rack systems. Older systems often pay little attention to contamination avoidance and a facility's lubrication work flow processes.

Figure 12.7: Minibulk storage indoors. Note approved transfer containers on upper shelf of cabinet (Ref. 3).

Figure 12.8: Modern lubrication work center (2010).

PERIODIC AUDITS

Periodic audits of lubrication practices are performed at leading-edge facilities, and many of these lube audits uncover unsatisfactory lubricant dispensing practices. Audit findings and recommendations often involve detection of oil contamination in sump-lubricated equipment (Figure 12.1) and the labeling of points to be lubricated.

Regrettably, some plants persist in arbitrarily "standardizing" on less than optimum oil and grease formulations. Many are employing incorrect regreasing practices on the millions of electric motors used to drive process pumps. Superior plants experience as few as 14 motor bearing replacements per 1,000 motors per year; average plants replace 156 motor bearing replacements per 1,000 motors per year (Ref. 1). The statistics of less-than-average plants are far worse.

Suffice it to say that being unaware of best available lubrication management and practices can be expensive. These expenditures could be avoided by simply keeping out contaminants. Experience shows that periodic lube management audits are cost-effective and almost always point out areas of improvement.

What We Have Learned

- Correct outdoor storage of oil drums involves considerable forethought and constant attention.
- Storage under a shelter or indoor storage is much preferred over full exposure to the ambient environment.
- Water and particulate contaminants can greatly impair bearing life. They must be kept away from pump bearings.
- Drum vent provisions are needed, but open vents that allow water and dirt ingress are unacceptable.
- Transfer containers should be purposefully designed and be kept immaculately clean.
- Periodic lube management audits are recommended.

References

1. Bloch, Heinz P.; *Machinery Reliability Improvement*, 3rd Edition (1998), Gulf Publishing Company, Houston, TX. (ISBN 0-88415-661-3).
2. SKF USA, Inc., Kulpsville, PA, General Catalog 4000 US, 1991.
3. Fluid Defense Systems, Montgomery, IL.

Chapter 13

Pump Condition Monitoring:

Pump Vibration, Rotor Balance, and Effect on Bearing Life

Process pump user-operators often want simple rules of thumb to determine maximum allowable vibration. Of course, rules of thumb should not be confused with statistical proof. Many times, general experience and common sense are of greater value than statistics. Essentially, this chapter deals with a few experience-based observations on the issue of condition monitoring for process pumps (Ref. 1).

VIBRATION AND ITS EFFECT ON BEARING LIFE

Pumps, like all rotating machines, vibrate to some extent because of the response from excitation forces, such as residual rotor unbalance, turbulent liquid flow, pressure pulsations, cavitation, and pump wear. The magnitude of vibration will also be amplified as flows deviate from best efficiency (Figure 13.1) and as the vibration frequency approaches the resonant frequency of a major pump, foundation, and/or connected piping.

Vibration from the running pump is often transmitted to the nonrunning (standby) pump. This transmitted vibration tends to wipe off the oil film on the bearings of nonrunning pumps, causing metal-to-metal contact. Bearing degradation then shows up when the standby pump is put in service. Degradation is often reduced by switching or alternating from the "A" pump to the "B" pump on a 4- to 6-week basis.

Several published observations on pump vibration and its effect on bearing life lead to a plot (Figure 13.2) that probably brackets 90% of all process pumps.

No two predictions are the same. Yet, Figures 13.2 and 13.3 illustrate the same point; vibration excursions tell a story and reflect the condition of a process pump. The root cause may be hydraulic and temporary; it could relate to flow disturbances that vary with the flow rate. Or, the root cause could be related to deficiencies in one or more mechanical components. Component imbalance or bearing defects may be causing vibration. Either

Figure 13.1:
Prominent pump
standards recognize
pump vibration
magnitude is not
uniform over the
entire flow range
(Source: API-610,
8th Edition, 1996).

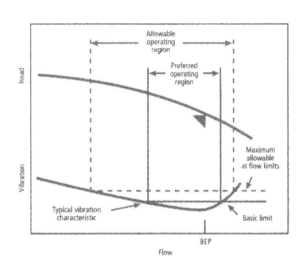

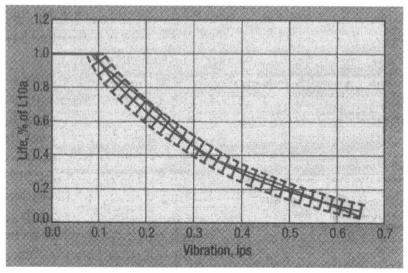

Figure 13.2: How pump vibration affects bearing life (Ref. 2).

way, vibration reduces bearing life in accordance with Figure 13.2.

The absolute value of vibration is not necessarily as important as the suddenness of a vibration increase. As an example, if pump vibration had been around 0.1 ips (2.5 mm/s) for the past 2 years and had increased to 0.3 ips (7.5 mm/s) in a single day, then we could consider this a more serious event than vibration that started at 0.3 ips and then gradually increased to 0.4 ips (10 mm/s) in the span of 12 months.

Figure 13.3 relates (conservatively) how an increase in overall vibration resulting from bearing-internal deterioration will shorten bearing life and by what percentage. It can be reasoned that vibratory activity adds to the normal bearing load. Bearing manufacturers report that rolling element bearing life varies exponentially with load. For a typical ball bearing, the exponent is 3; therefore, a two-fold load reduces bearing life by a factor of $2^3 = 8$. The bearing life is then only 12% of what it would have been at normal load conditions. A compelling case is thus made for keeping vibration low.

For most pump reliability improvement professionals, the issues of interest are not whether or not the pump vibrates but instead include the following:

* If the amplitude and/or frequency of the vibration is sufficient to cause actual or perceived damage to any of the pump components

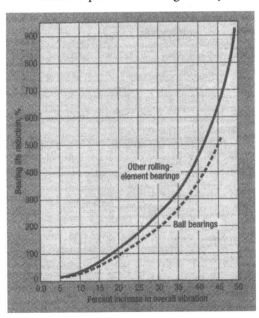

Figure 13.3: Bearing degradation shows up as vibration and can be related to life reduction.

- If the vibration is a symptom of some other damaging phenomenon happening within the pump

- If the relationship between vibration severity and bearing life can be quantified with a reasonable degree of accuracy

Various industry organizations, such as the Hydraulic Institute (HI, in ANSI/HI 9.6.4, Ref. 3), and the American Petroleum Institute (API, in its Standard API-610) have set pump vibration limits for general guidance. All are reaching back to the experience of individual reliability professionals and multinational pump user companies that had implemented daily machinery condition reviews (monitoring and surveillance) decades ago.

Some of these professionals had assisted operating personnel by listing acceptable, reportable, and mandatory shut-down levels of vibration. These levels represent experience-based values that rely on the bearing life versus vibration approximations for general-purpose machinery. They have been widely published since the late 1940s as allowable vibration velocities.

Traditionally, 0.35 ips (~9 mm/s) was given as a maximum allowable vibration velocity for "total all pass" (overall) readings taken on bearing caps or housings. However, machinery vibration and its measurement are complex matters and may require some clarification. Typical considerations might include the following:

1.   Vibration can be measured and/or analyzed by using units of displacement, velocity, or acceleration severity to evaluate the health of the machine. As stated earlier, the primary or traditional measure of vibration used by industry today is velocity. Because most pump users use this parameter, comparisons are made easier.

2.   Either the "total all-pass" or the "filtered" frequency can be used. Most industry specifications and standards use "total all-pass" vibration values to identify problem pumps. Filtered values are reserved for determining where the vibration originates. This latter determination is generally called "vibration analysis."

3.   Root mean square (RMS) as well as peak-to-peak values are sometimes measured or specified. The HI has chosen RMS acceptance limit values. HI recognizes that most vibration instruments actually measure vibration in RMS terms and then calculate peak-to-peak values, if re-

quired. API, however, generally refers to peak-to-peak readings. RMS values are roughly 0.7 multiplied by peak measured values.

However, this relationship applies only to vibration consisting of a single sinusoidal waveform. For more complex waveforms, this conversion does not yield correct results.

4. The acceptable vibration amplitude (as-new vs. post-repair levels) may have to be specified for a particular application. Acceptance limits will change along with overall pump power and flow rate regions. The Hydraulic Institute and International Standards Organization (ISO) base acceptable vibration limits on pump type and power level, whereas API gives different limits for the "preferred" and "allowable" operating regions. (See vibration acceptance limits, below, for "as-new" acceptance values.)

5. It is difficult to predefine how factory test stand vibration measurements should compare with field (at site) values. The exact effects of foundation stiffness/grouting are difficult to predict. Generally, the stiffness of a field pump foundation is much higher than the stiffness found on a factory test stand, especially if the pump base plate is grouted. That is why the Hydraulic Institute vibration standard allows higher test stand values (up to twice field values).

For vertical turbine pump installations, it is especially important to know the actual foundation stiffness to avoid high vibration from operation at a structural resonance frequency.

6. Cataloging how much the vibration amplitude and/or frequency has changed over the life of the machine is important. This is called "trending." It is especially helpful to have an as-new vibration signature taken and kept on file for future comparison.

7. Location of vibration measurements: On a typical horizontal process pump, vibration readings are taken in the x, y, and z (axial) directions.

Horizontal and vertical dry pit pump vibration measurements are normally taken on or near the outer, or uppermost, bearing) in the horizontal, vertical, and axial planes, with the maximum value used for acceptance.

Vertical turbine pump vibration measurements are taken at the top or bottom of the motor. Probes should not be located on flexible panels, walls, or motor end covers.

MONITORING METHODS DIFFER

Some data gathering methods employ shock pulse monitoring (SPM) and, like all other methods, may or may not forward the results by wireless means. In simple terms, the SPM method detects the development of a mechanical shock wave caused by the impact of two masses. At the exact instant of impact, molecular contact occurs and a compression (shock) wave develops in each mass. The SPM method is based on the events occurring in the mass during the extremely short time period after the first particles of the colliding bodies come in contact. This time period is so short that no detectable deformation of the material has yet occurred. The molecular contact produces vastly increased particle acceleration at the impact point. The severity of these impacts can be plotted, trended, and displayed.

There is also temperature monitoring. Suppose a pump bearing housing operates at 170°F (~77°C). Most people can place an index finger on such a bearing housing for about 5 seconds before the pain becomes too intense. However, 170°F (~77°C) is not excessive for process pumps (see Chapter 8). Proper pump surveillance calls for measurements with either a surface pyrometer or a hand-held noncontacting infrared heat sensing instrument (thermal gun).

Under no circumstances should the bearing housing be doused with water. Such cooling would probably cause the bearing outer rings to be cooled. Because metal shrinks upon being cooled, the already small bearing-internal clearances would be reduced to the point of being excessively preloaded. Bearing life would be curtailed by the method thought to extend it.

VIBRATION ACCEPTANCE LIMITS

Hydraulic Institute Standard ANSI/HI 9.6.4 presents the generally accepted allowable pump "field" vibration values for various pump types (see Table 13.1). The standard is based on RMS total or all-pass vibration values. The standard states that factory or laboratory values can be as much as twice these field limits, depending on the rigidity of the test stand. The ANSI/ASME B73 standard accepts two times the HI 9.6.4 values for factory tests performed on chemical end suction pumps (Ref. 3). HI includes the API-610 values for end suction refinery pumps (in RMS terms); the API-610

Table 13.1: Allowable field-installed vibration values for pumps (Ref. 3).

PUMP TYPE	LESS THAN POWER (HP)	VIBRATION RMS, IN/SEC	GREATER THAN POWER (HP)	VIBRATION RMS, IN/SEC
End Suction ANSIB73	20	.12	100	.18
Vertical Inline, Separately-Coupled, per ANSI B73.2	20	.12	100	.18
End Suction & Vertical Inline Close-Coupled	20	.14	100	.21
End Suction, Frame-Mounted	20	.14	100	.21
End Suction, API-610, Preferred Operation Region (POR)	All	.12	All	.12
End Suction, API-610, Allowable Operation Region (AOR)	All	.16	All	.16
End Suction, Paper Stock	10	.14	200	.21
End Suct. Solids Handling – Horizontal	10	.22	400	.31
End Suct. Solids Handling – Vertical	10	.26	400	.34
End Suct. Hard Metal/Rubber-Lined, Horizontal & Vertical	10	.30	100	.40
Between Bearings, Single & Multistage	20	.12	200	.22
Vertical Turbine Pump (VTP)	100	.24	1000	.28
VTP, Mixed Flow, Propeller, Short Set	100	.2	3000	.28

document requires that these acceptance values be demonstrated on the factory test stand.

The HI Standard also states that stipulated values only apply to pumps operating under good field conditions. Good field conditions are defined as follows:

1. Adequate net positive suction head (NPSH) margin
2. Operation within the pump's preferred operating region—typically 70% and 120% of BEP (Table 13.3 only lists the constant values required for low and high pump power ratings. The acceptable vibration, between the low- and high-power values, varies linearly with power on a semi-log graph)
3. Proper pump/driver shaft (coupling) alignment
4. Pump intake must conform to ANSI/HI 9.8 ("Pump Intake Design")

It should also be noted that the acceptable vibration values for slurry and vertical turbine pumps are about double the values given for horizontal clean liquid pumps.

Once a pump is accepted and commissioned, somewhat higher total (all-pass) vibration values are usually accepted before additional follow-up and analysis are deemed appropriate. As a general rule, repair follow-up is recommended if vibration levels increase to twice the "field" acceptance limits (or initial actual readings).

CAUSES OF EXCESSIVE VIBRATION

Once a pump has been determined to have a high "total or all-pass" vibration level, the next step is to identify the cause. This would be the time to obtain a filtered vibration analysis and to look for predominance of one of many frequencies in the spectrum. Table 13.2 illustrates several predominant frequencies, although it is providing a narrow overview at best.

Along those lines, the first step in the analysis should be to capture, and then evaluate, the multiples of pump running speed (Table 13.2). A graphic display would often be called a "filtered" velocity plot or frequency spectrum. Actual analysis can point to the following possible causes:

1. Rotor unbalance (new residual impeller/rotor unbalance or unbalance caused by impeller metal removal—wear)

2. Shaft (coupling) misalignment
3. Liquid turbulence resulting from operation too far away from the pump best efficiency flow rate
4. Cavitation resulting from insufficient NPSH margin
5. Pressure pulsations from impeller vane—casing tongue (cutwater) interaction in high discharge energy pumps

Table 13.2: Sources of specific vibration excitations.

FREQUENCY	SOURCE
0.1 x Running Speed	Diffuser Stall
0.8 x Running Speed	Impeller Stall (Recirculation)
1 x Running Speed	Unbalance or Bent Shaft
2 x Running Speed	Misalignment
Number of Vanes x Running Speed	Vane/Volute Gap and Cavitation

Other possible causes of vibration may be more complex to analyze. Among these are: Operating speed close to mechanical or hydraulic resonant frequencies of a major pump, foundation, or pipe component. This is of special importance with large multistage horizontal and long vertical pumps. A margin of safety should be provided between rotor and/or structural natural frequencies and operating speed. Typical margins are 15-25%. Vibration amplification will generally be greater than 2.5 times at a resonant frequency.

Vibration/resonance events to be evaluated on pumps include rotor lateral vibration and structural lateral vibration—rather common with long-shafted vertical pumps.

Poor pump suction or discharge piping can also cause increased vibration, normally by either increased cavitation or turbulent flow in the pump. Pump operating speed or vane pass frequencies could excite a piping structural or hydraulic resonance. (For additional comments on piping vibration see Chapter 4.)

Bearing wear will usually show up in the vibration spectrum. Rolling element bearings have distinct vibration signatures based on the number of bearing balls or rollers. Recall, however, that monitoring deterioration of plastic bearing cages would require highly sophisticated monitoring techniques. This is one of the reasons why plastic cages can be used in pump bearings only after all relevant factors are taken into account.

Opening up of impeller wear ring clearances is primarily shown in performance measurements (Ref. 1). This wear can reduce the NPSHr margin and shift the pump operating flow point. It will not, however, generate vibration activity that is easily monitored by conventional means.

ROTOR BALANCING

All impellers, irrespective of their operational speed, should be dynamically balanced ("spin-balanced") before installation, either single or two plane. Two-plane balance is required for a wide impeller, typically when the impeller width is greater than 17% of the impeller diameter. ISO balance criteria are usually invoked, and an experienced balance shop will know them well.

Dynamic balancing of the three major rotating pump components, shaft, impeller, and coupling, will increase mechanical seal and bearing life. All couplings in the weight or size ranges found in a modern refinery should be balanced if they are part of a conscientious and reliability-focused pump failure reduction program.

Of course, if a facility is willing to remain repair-focused, then it can continue to plod along with "business as usual." Still, reliability-focused plants agree: Large couplings that cannot be balanced have no place in the majority of their process pumps.

The preferred procedure for process pumps in reliability-focused installations is to balance the impeller and coupling independently and then to balance the impeller and coupling on the shaft as a single unit. Another method is to balance the entire pump rotor as an assembled unit and to do so one time only. That might be a bit problematic at locations that will subsequently go through repair cycles while trying to omit full rotor balance. Often, more problems are caused in successively disassembling and reassembling than would be caused by diligently balancing each individual component. For multistage pump rotors (both horizontal and vertical), individual component balance is generally preferred.

The static (single plane) balance force is always the more important of the two forces—static and dynamic ("couple force"). If balancing of individual rotor components is chosen, then it is best to use a tighter tolerance for the static (single plane) force. In theory, if the static force is removed from each part, then there should be little dynamic (couple) force remaining in the rotor itself.

For impellers operating at 1,800 rpm or less, the ISO 1940 G6.3 tolerance is acceptable. For 3,600 to 1,800 rpm, the ISO G2.5 rule is better. Both are displayed on balance tolerance nomograms for small and large machinery rotors. Generally, tighter balance tolerances (G1.0) are not warranted unless the balancing facility has modern, automated balancing equipment that will achieve these results without adding much time and effort.

Using older balancing equipment may make it difficult and unnecessarily costly to obtain and duplicate the G1.0 quality. Also, factory vibration tests have, at best, shown insignificant reductions in pump vibration with this tighter balance grade.

That, however, is not the point. Instead, let us realize that relatively tight balance tolerances or good grades of balance are obtained on automated balancing machines just as quickly as would more liberal, less precise, balance specifications. Using an analogy, why allow bottles of medicine to contain between 99 and 101 tablets when modern filling machinery can guarantee to deliver precisely 100 tablets per bottle? Surely, a serious and reliability-focused user-consumer will insist on products with consistently high quality.

Balancing machine sensitivity must be adequate for the part to be balanced. This means that the machine should be capable of measuring unbalance levels to one-tenth of the maximum residual unbalance allowed by the balance quality grade selected for the component being balanced.

Rotating assembly balance is recommended whenever practical and if the tighter quality grades, G2.5 or G1.0, are desired. Special care must be taken to ensure that keys and keyways in balancing arbors are dimensionally identical to those in the assembled rotor. Impellers must have an interference fit with the shaft when G1.0 balance is desired. Although looseness between impeller hub and shaft (or balance machine arbor) is allowed for the lesser balance grades, it should not exceed the values given in Table 13.3 for grades G2.5 or G6.3:

Table 13.3: Maximum diametral hub looseness.

Impeller Hub	Maximum Diametral Looseness	
Bore	< 1,800 rpm	1,800-3,600 rpm
0 – 1.499 in.	.0015 in.	.0015 in.
1.5 – 1.999 in.	.0020 in.	.0015 in.
2.0 in. and larger	.0025 in.	.0015 in.

WHAT WE HAVE LEARNED

• Initial guidance on allowable pump vibration is clearly available from the hundreds of articles and dozens of books that have been published in the decades since 1960. Up-to-date summaries are contained in Ref. 2 and other modern texts.

• Because elevated vibration increases the forces acting on bearings, and because bearing life is related to bearing load, higher vibration will reduce bearing life. The rules of thumb and empirical relationships express these guidelines with sufficient accuracy for general purpose equipment.

• Modern data collectors and condition analyzers are available from several competent vendor-manufacturers. Also, many models are operating with wireless connections, whereas others are hand-held or hardwired (Ref. 4). Each has its advantages and these must be considered on a case-by-case basis.

• Because rotor unbalance will lead to increased vibration, good rotor balance is essential. The issue of balance grade is moot in the many instances in which modern, often fully automated, balancing machines are readily available. These balancing machines will achieve excellent equipment rotor balance as quickly and effectively as not-so-excellent balance.

• Bearing life is related to shaft misalignment and force transmission across couplings. These affect vibration severity (Ref. 5). Although rules of thumb are not absolutes, their judicious application makes far more sense than rather simplistic requests to "prove it to me."

The issue is about risk and the mitigation of risk. A fitting analogy deals with automobiles, in which reasonable people know that driving on worn tires will put the passengers at greater risk than driving on new tires.

Likewise, the issues of vibration and shaft misalignment are intuitively evident to most of us. We should be satisfied with rules of thumb and empirical data where they appeal to common sense. They certainly do in this instance.

Do not compromise safety and reliability. Keep process pump vibration low.

References

1. Beebe, Ray S.; "Machine Condition Monitoring," *Engineering Handbook* (2001), MCM Consultants, Monash University, Gippsland, Australia.
2. Bloch, Heinz P., and Alan Budris; *Pump User's Handbook,* 3rd Edition (2010), The Fairmont Press, Lilburn, GA. (ISBN 0-88173-627-9).
3. ANSI/HI Standard 9.6.4; Hydraulic Institute, Parsippany, NJ (2001).
4. Bloch, Heinz, Paul Lahr, and Donald Hyatt; "Development of an advanced electronically optimized variable high-speed centrifugal pump," Proceedings of Pump Congress, Karlsruhe, Germany, October 4-6, 1988.
5. Berry, Douglas L.; "Vibration vs. Bearing Life Increase," *Reliability*, December 1995.

Chapter 14

Drivers, Couplings, and Alignment

Recall, please, that this text deals with pump-related issues that are often overlooked. Drivers, couplings, and the accuracy of shaft alignment are pump-related and affect availability and operating life. They deserve our attention.

DRIVER SELECTION

Driver selection is based on criteria that involve many different considerations, including commercial and in-plant power grids, substations, and so on. Types of winding insulation are specified by the design contractor, but the power demand of a particular process pump goes up during episodes of pump overload or as pump internals wear. These load increases are not always addressed in the motor sizes being ordered.

For a pump requiring 125-hp input, buying a 150-hp motor with a service factor (S.F.) of 1.0 will provide a longer ultimate motor life than buying a 120-hp motor with an S.F. of 1.15. Proper motor selection should also take into account temperature profiles and installation geography (i.e., elevation above sea level). All can influence motor efficiency and winding life.

Motor lubrication either should be specified by the purchaser or, alternatively, be discussed with the motor vendor. This discussion and approval step is especially important with oil mist lubrication, where a special type of insulating tape must be used in the junction box (Ref. 1). Other than that, oil mist lubrication on motors has been highly successful since the mid-1960s and, on rolling element bearings, is much preferred over other lubrication methods. The ingress of pure oil mist into a modern electric motor will not adversely affect it. The slightly pressurized mist environment may, in fact, keep out airborne dirt.

Approval and disclosure of the path traveled by grease through a motor bearing is also important. Many times, electric motor bearings are not arranged for effective relubrication. The intent of shielded bearings (whereby the replenishing grease actually enters the rolling elements through capillary action and not through direct pressurization!) is sometimes poorly understood by all parties. Application of undue pressure to a bearing shield can cause the rolling elements to scrape on the shield. Injecting grease at excessive pressure often hastens motor bearing failures (Ref. 2).

COUPLING SELECTION AND INSTALLATION

The lowest initial cost coupling is rarely the best choice once life-cycle cost and safety risk are put into the equation. Process pump couplings should have an S.F. of 2 or 2.5. Elastomeric couplings designed with the flexible element twisting or pulling should not be used for large pumps. The following number of factors come into play here:

•   Toroidal ("tire-type") flexing elements will exert an axial pulling force on driving and driven bearings. Also, they are difficult or impossible to balance.

•   Polyurethane flexing elements perform poorly in concentrated acid, benzol, toluene, steam, and certain other environments.

•   Polyisoprene flexing elements do poorly in gasoline, hydraulic fluids, sunlight (aging), silicate, and certain other environments.

Gear couplings require grease replenishment and are certainly more maintenance-intensive than nonlubricated disc pack couplings. If gear couplings are selected, then they should be on the plant's preventive maintenance schedule. Only approved coupling greases should be used for periodic replenishment. Realize that standard and multipurpose greases are unsuitable because their oil and soap constituents get "centrifuged apart" at typical pump coupling peripheral speeds.

That leaves a reliability-focused user with a nonlubricated disc pack (Figure 14.1) or other variants of alloy steel membrane couplings as the preferred choice. The design must be such that the spacer piece or center member is captured. In the unlikely event of a disc pack failure, a "captured center member" will not leave the space between the two coupling hubs. Many brands of cheap couplings do not have captured center members.

INSTALLATION AND REMOVAL

Before installing a coupling, examine it for adequacy of puller holes or other means of future hub removal. The coupling in Figure 14.2 was mistreated at disassembly because no thought had been given to future removal.

For parallel pump shafts with keyways, use 0.0-0.0005 inches (0.0-

Figure 14.1: A typical captured center member disc pack coupling for process pumps.

Figure 14.2: Coupling hub damaged with severe hammer blows at disassembly.

0.012 mm) total shaft interference. Use one of several available thermal dilation methods (heat treatment oven, superheated steam, or electric induction
heater) to mount hubs on shafts.

Disallow loose-fitting keys for coupling hubs because they tend to
cause fretting damage at shaft surfaces. During rebuilding or repair, allocate
time needed for hand-fitting keys; they should fit snugly in keyway. On all
replacement shafts, machine radiused keyways and modify the keys to match
the radius contour.

ALIGNMENT AND QUALITY CRITERIA

Pump base plates should incorporate means of adjusting pumps and
drivers for proper shaft alignment. Jacking bolts for horizontal movement
in two directions were discussed in Chapter 3 and are shown in Figure 14.3.
Shaft alignment work can be carried out with the bearing housing support
brackets bolted up securely in this instance. Here, the pump's two large size
main support pedestals and its bearing support bracket can reasonably be
expected to operate at the same temperature.

Figure 14.3: A quality process pump during factory assembly. Note alignment
jacking bolts on pump supports and on motor (Source: Emile Egger & Cie.,
Cressier, NE, Switzerland).

But, suppose a pump casing had mounting feet integrally cast with the pump casing (Figure 14.4), and these integral mounting feet would be bolted to the base plate. In that case, the thermal growth of the (hot) pump might differ from that of the (somewhat colder) ambient air-flooded bearing support bracket. The bracket would have to be unbolted while shaft alignment work is in progress. The bracket needs to be reconnected after the pump has been started and has reached temperature equilibrium.

Note that the pump model shown in Figure 14.5 does not incorporate a bearing support bracket. Designing a sturdy pump without the need for a bearing support bracket circumvents the issue mentioned in the previous

Figure 14.4: Foot-mounted ANSI pump on a conventional base plate. On other than ambient temperature services, the bearing support bracket must be left unbolted while the pump grows up or down to reach thermal equilibrium at its operating temperature.

Figure 14.5: This medium-size pump is designed without a bearing support bracket. It incorporates motor jacking bolts and is shipped fully aligned on an epoxy-filled base plate (Ref. 3).

paragraphs. (Do not overlook that a coupling guard must be installed before allowing this pump—or any other pump—to be put into service. An epoxy prefilled base plate with motor jacking provisions is shown on this fully mounted pump set.)

Several different alignment methods exist; the reverse indicator method setup of Figure 14.6 is still widely used. Reverse dial indicator alignment is sound and appropriate. However, it must be properly executed and due attention must be given to indicator bracket sag. As is the case with all pump alignment methods, the task encompasses cold-aligning with offsets to compensate for equipment thermal growth. Alignment work crews must be well trained and conscientious.

The reverse dial indicator method forms the basis of its successor technology: modern laser alignment techniques. Because laser alignment is both fast and precise, it has become the preferred method. It is used extensively by best practices companies.

CONSEQUENCES OF MISALIGNMENT

Pump misalignment has serious consequences. Among other shortcomings, misalignment may be forcing standard bearings to operate at an unintended angle. The simplest approximation of misalignment is ob-

Figure 14.6: Reverse dial indicator alignment setup using low-sag tubular indicator mounting brackets (Ref. 3).

tained by measuring the parallel offset between a pump shaft and the shaft of its driver. Also, the distance between the two shaft ends is measured. The tangent is a trigonometric function; it is obtained by dividing the shaft offset (inches or mm) by the distance between shaft ends (inches or mm), often labeled DBSE.

To achieve the rated bearing life of 100%, Ref. 4 recommends keeping the tangent of the misalignment angle at 0.001 and lower. This reference describes tests carried out on a variety of bearings, and for most of these, the effect of misalignment on life is described by the boundaries of the two curves in Figure 14.7. This graphic describes that once the tangent exceeds 0.006, rolling element bearing service life drops into the 10-20% range. However, aiming for a tangent value of 0.001 is rather generous compared with the considerably more stringent tangents found at some best-of-class pump user companies. Best-of-class practice is not to allow shaft misalignments in excess of 0.5 mils per inch (0.0005 mm per mm) of shaft separation. This rather severe guideline makes allowance for angular misalignment that may exist in addition to parallel offset.

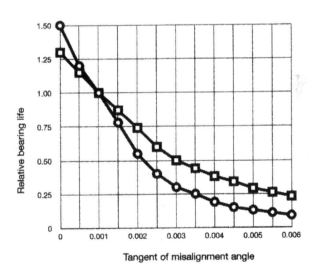

Figure 14.7: How misalignment shortens service life of typical bearings. Most bearings fit somewhere between the two curve boundaries. At tangents below 0.001, bearing life is here assumed to exceed a relative rating of 1 (Ref. 4).

THERMAL RISE AND PREDEFINITION OF GROWTH

We assume that process pumps are driven by electric motors or small steam turbines. As these drivers start up, they also warm up. Of course they "grow" as the temperature increases and this thermal growth must be taken into account when aligning driver and driven shafts.

Because a typical electric motor driver does not get too hot, its thermal growth can be neglected. The same is true for foot-mounted process pumps with product temperatures up to about 200 °F (~93°C). On foot-mounted pumps with process temperatures in excess of 200°F the cold offset (in inches) can be calculated from

$$\Delta H = (0.000006) \ (H) \ (\Delta T)$$

In this formula, ΔH is the height difference below which the pump shaft should be set in the cold condition. H is the height or distance from the bottom of the feet of the pump to its centerline, and ΔT is the temperature (°F) of the pumped fluid after subtracting 200.

Here is an example for a pump with a distance from the bottom of the feet to the shaft centerline equal to 18 inches and a process temperature of 430°F:

$$\Delta H = (0.000006) \ (18) \ (430 - 200) = 0.025 \text{ inches}$$

This pump is expected to grow about 0.025 inches as a result of being heated 230° hotter than the driver. Of course, this is only an approximation, but it will work.

On American Petroleum Institute (API)-style centerline-mounted pumps (Figure 13.5) with process temperatures in excess of 200°F the cold offset can be calculated from

$$\Delta H = (0.000002) \ (H) \ (\Delta T)$$

In this formula, ΔH is the height difference below which the pump shaft should be set in the cold condition. H is the height of the pedestal (usually the distance from the base plate to the pump centerline) and ΔT is the temperature of the pumped fluid after subtracting 200.

Here is an example for a pump with a pedestal height of 21 inches and a process temperature of 740°F:

$$\Delta H = (0.000002) \ (21) \ (740 - 200) = .0022 \ \text{inches}$$

This pump is expected to grow about 0.022 inches because of being heated 540° hotter than the driver. This, too, is only an approximation, but it will work.

Aim not to allow shaft misalignment to exceed the limits plotted by competent alignment service providers. Use 0.5 mils per inch (0.0005 mm per mm) of distance between shaft ends (DBSE) as the maximum allowable shaft centerline offset.

In the many decades since 1950, a body of literature has sought to quantify the adverse effects of misalignment on rotating machinery. Figure 14.8 is simply one of such numerous efforts put forth by highly experienced observers.

Of course, Figure 14.8 is somewhat general and many variables determine its precise accuracy. However, it should be used to justify good alignment practices and support the contention that pump misalignment causes bearing overloads and the loss of reliability. Such overloads also create frictional resistance that consumes power. Given the sizeable benefit-to-cost ratio, precision alignment is a key ingredient to optimizing the life-cycle cost of process pumps.

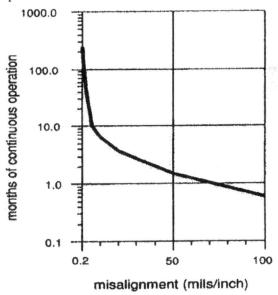

Figure 14.8: Months of continuous operating life versus misalignment (Ref. 5).

WHAT WE HAVE LEARNED

- Specify and purchase pump sets with jacking arrangements at driver and pump.

- Grease and oil mist lubricated drivers require proof that the manufacturer understands both user requirements and relevant technology

- Although not directly considered part of the pump, couplings must be selected with thought given to installation, removal, safety, periodic maintenance, and future performance.

- A size-on-size or 0.0005-inch shaft coupling bore-to-shaft interference fit will minimize shaft fretting and shaft weakening. Careful procedures are needed; they should take into account that this represents a close clearance.

- Keys should be hand-fitted to shaft keyways.

- Coupling hubs must incorporate removal provisions such as puller holes.

- A reverse dial indicator or laser alignment methods should be used.

- Do not allow shaft misalignments of more than 0.5 mils per inch (0.0005 mm per mm) of shaft separation (shaft separation is also called DBSE).

References

1. Bloch, Heinz P., and Abdus Shamim; *Oil Mist Lubrication: Practical Applications* (1998), The Fairmont Press, Lilburn, GA. (ISBN 0-88173-256-7).
2. Bloch, Heinz P.; *Practical Lubrication for Industrial Facilities*, 2nd Edition (2009), The Fairmont Press, Lilburn, GA. (ISBN 0-88173-579-5).
3. Bloch, Heinz P. and Alan Budris; *Pump User's Handbook*, 3rd Edition (2010), The Fairmont Press, Lilburn, GA. (ISBN 0-88173-627-9).
4. Leibensperger, R.L.; "Look beyond catalog ratings," *Machine Design*, April 3, 1975.
5. Piotrowski, John; *Shaft Alignment Handbook*, 3rd Edition (2006), MarcelDekker, New York, NY. (ISBN 978-157444-7217).

Chapter 15

Fits, Dimensions, and Related Misunderstandings

No pump has ever been built (nor will ever be built) for simultaneously operating at the extremes of speed, head developed, temperature allowed, limit of dimensional fits, tolerances, pipe-flange induced stress, marginal lubrication, and so forth. Allowing every relevant dimension or parameter to be at their absolute limits at the same time will surely cause pumps to fail prematurely and will put others at risk.

Centrifugal process pumps in U.S. oil refineries and other plants typically reach mean-times-between failures (MTBFs) ranging from barely 2 years to as much as 10 years (Ref. 1). It can be reasoned that the MTBF of these simple machines is largely influenced by issues or parameters that "someone" either misunderstood or chose to disregard. We have found that many pump failure incidents can be avoided by acting on the information contained in the checklist portion of this chapter.

LOW INCREMENTAL COST OF BETTER PUMPS

On average, a significantly more reliable pump costs 20% more than its "bare necessity" counterpart. Upgrading during the next repair event is likewise estimated to cost about 20-30% of a "bare necessity" repair.

Independent estimates from at least two petrochemical corporations hold pump issues responsible for one fire per 1,000 pump repair events. With the average pump fire causing several million dollars worth of damage, achieving fire risk reduction by component upgrading will be worth considerable money. Smart users are, therefore, factoring the imputed value of fire avoidance into their cost justifications.

In the late 1990s and early 2000s, the average repair cost of a defective refinery pump was approximately $11,000—a figure that included burden and overhead. (Ref. 1). By any measure, implementation of the average improvement item costs less than $2,000 and is worth every penny of it. Our checklist of implementation items lists "things to consider" when pursuing reliability improvement at the point of initial purchase. The same list would apply while carrying out repairs on existing pumps.

A pertinent pump checklist is filled with items, issues, and procedures known to best practices performers. Many more detailed descriptions can be found in earlier chapters of this text on pump wisdom; others can be found in the reference literature.

Our comprehensive checklist starts with cooling issues, followed by bearing topics, lubrication issues, and then mechanical seal issues. Many of these items or topics overlap (i.e., a cooling topic overlaps with certain lubrication topics, and these in turn, overlap with bearing protector and leakage prevention issues, etc.).

Just remember that checklists are not intended to take the place of the rigorous explanations that can be found in narrative texts.

PUMP PEDESTALS AND BEARING HOUSINGS SHOULD NOT BE WATER-COOLED

*   Do not allow pedestal cooling of centrifugal pumps, regardless of process fluid or pumping temperature. Pedestal cooling is inefficient and corrosion starting from the inside of pedestals has caused massive support failures. If you find a pump with water-cooled pedestals, do the following instead:
    (a) Calculate and accommodate thermal growth by appropriate cold alignment offset, per Chapter 14.
    (b) Consider using hot alignment verification measurements (Refs. 2 through 4).

*   Do not allow jacketed cooling water application on bearing housings equipped with rolling element bearings.
    (a) Note that water surrounding only the bearing outer ring will sometimes cause bearing-internal clearances to vanish, leading to excessive temperatures, lube distress, and premature bearing failure.
    (b) Forced-air cooled "finned" bearing housings are acceptable because the bearing outer ring has sufficient clearance in the housing bore so as not to be excessively preloaded by the cooling effect of the

air. However, even air-flow cooling may simply not be necessary if superior synthetic oil formulations are used.

- Understand and accept the well-documented fact that, on pump bearing housings equipped with rolling element bearings, it is possible to delete cooling of any kind (Ref. 5). For cost and reliability-optimized results:

    (a) Simply change over to the correct synthetic lube type and lube formulation. For best results, the lubricant will have to contain proprietary additives.

    (b) Use a suitably formulated synthetic lubricant that, at operating temperature, exhibits viscosity characteristics needed for the oil application method used in that particular bearing housing.

    (c) Recognize that cooling water coils immersed in the oil sump may cool not only the oil but also the air floating above the oil level. The resulting moisture condensation can cause serious oil degradation. This is one of several reasons for avoiding water cooling of pump bearing housings.

    (d) Recognize that stuffing box cooling is generally ineffective. If mechanical seals need cooling, then investigate alternative seal flush injection and external cooling methods. (In the 1960s, it was proven that with stuffing box cooling the face temperature of mechanical seals is decreased by only about 2°F, i.e., little more than 1°C).

- Cooling water may still be needed for effective temperature control in sleeve bearing applications (and these are not generally addressed in this text). Again, it should be noted that excessively cold cooling water will often cause moisture condensation.

    (a) Even trace amounts of water may greatly lower the ability of lubricants to adequately protect the bearing.

    (b) On sleeve bearings, close temperature control is far more important than on rolling element bearings.

SUMMARY OF BEARING-RELATED ISSUES

- Do not use filling-notch bearings in centrifugal pumps. The inevitable axial loading will cause rapid bearing degradation. Replace filling-notch bearings with dimensionally identical deep groove ("Conrad-type") bearings.

- All bearings will deform under load.
  - (a) On thrust bearings that allow load action in both directions, deformation of the loaded side could result in excessive looseness and, hence, skidding of the unloaded side.
  - (b) This skidding may result in serious heat generation and thinning-out of the oil film. Metal-to-metal contact will now destroy the bearing.
  - (c) Always select bearing configurations that limit or preclude skidding. The better bearing manufacturers have application engineering departments that can advise suitable replacement bearing upgrades. Work with these manufacturers and be prepared to pay a little more for their bearings. It will be well worth it in the long run!

- The API-610 recommended combination of two back-to-back mounted 40-degree angular contact bearings is not always the best choice for a particular pump application.
  - (a) Matched sets of well-engineered 40-degree and 15-degree angular contact bearings are designed to avoid or greatly reduce skidding. Although not a cure-all, they may thus be more appropriate for a given service application.
  - (b) On some pumps, sets of 29-degree or sets of 15-degree angular contact bearings should be used.

- Keep in mind the possibility of using a 9000-series thrust bearing together with a 7200- or 7300-series angular contact bearing in the thrust location of certain pumps.
  - (a) Seek factory or application engineering advice and understand best-of-class lubricant application method before proceeding.

- Thrust bearing axial float (i.e. the total amount of movement possible between thrust bearing outer ring and bearing housing end cap, should not exceed 0.002 inches [0.05 mm]). To comply with this recommendation, some hand fitting or shimming may be necessary. This will limit what might otherwise become potentially excessive bearing-internal acceleration forces.

- On certain ANSI pumps, consider replacing old-style double-row angular contact bearings (bearings with one inner and one outer ring) with newer, Series 5300UPG, double-row/double inner ring angular contact bearings.
  (a) Series 5300UPG bearings have two brass cages, one outer ring and two inner rings per bearing.
  (b) With series 5300UPG bearings, axial clamping of the two inner rings is needed; the advantage is that these newer double-row bearings resist skidding.

- Observe allowable assembly tolerances for rolling element bearings in pumps with single angular contact (Conrad-type) and double-row bearings:
  (a) Bore-to-shaft: 0.0002-0.0007 inches (0.005-0.018 mm) interference fit
  (b) Bearing outside diameter-to-housing fit: 0.0007-0.0015 inches (18-37 μ) loose fit

- Do not use bearings with the least-expensive plastic cages in process pumps expected to operate dependably for years:
  (a) Some plastic cages tend to be damaged unless highly controlled mounting temperatures are maintained.
  (b) Plastic cage degradation will not show up in conventional vibration data acquisition and analysis.
  (c) Superior high-performance plastic cages were developed in the early 2000s. Consider using them if the application engineering group of a competent bearing manufacturer can point to solid experience.

- Be certain to use only precision-ground, matched sets of thrust bearings in either back-to-back thrust or tandem thrust applications.
  (a) Matched bearings must be furnished by the same bearing manufacturer.
  (b) Verify precision grinding by observing that appropriate alphanumerics have been etched into the back (the wide shoulder) of the outer ring.

- Use radial bearings with C3 clearances in electric motors to accommodate thermal growth of the hotter-running bearing inner ring.
  (a) Note that modern polyurea greases ("EM" greases) are much preferred for electric motors.

• Investigate column bearing materials upgrade options on vertical deep-well pumps (earlier shown in Figure 2.3) and compare available high-performance (HP) polymers.

   (a) Understand all physical properties and the intended service before picking the right HP polymer bearing for the job.

• Next to oil-jet lubrication, pure (also called "dry-sump") oil mist applied in through-flow fashion per API-610 (8th and later editions) represents the most effective and technically viable lubrication and bearing protection method used by reliability-focused industry.

   (a) Dry-sump (pure) oil mist is also one of the most successful lube application methods for electric motor bearings.

   (b) Dry-sump (pure) oil mist is always applied to nonrunning standby pumps and drivers. It protects their bearing housings against the intrusion of airborne moisture and dust.

   (c) Taking into account all of the previous points, investing in oil mist systems will usually provide paybacks ranging from 8 to 30 months.

   (d) Closed oil mist systems have been in use since the 1960s. They do not allow stray mist to escape to the environment.

   (e) Oil mist is virtually maintenance-free.

• Oil-jet system retrofit options are easily engineered for new installations as well as upgrades and conversions.

   (a) A simple and economical inductive pump (i.e., a small pump with a free piston as its only moving part) can serve as the source of a continuous stream of pressurized lube oil.

   (b) Inductive pumps can be used in conjunction with a spin-on oil filter. The resulting clean stream of lubricant can be directed at the bearing rolling elements for optimum effect.

• Realize that oil ring lubrication rarely represents state of art. Oil rings are highly shaft alignment-sensitive and tend not to perform dependably if one or more of the following requirements are not observed:

   (a) Unless the shaft system is absolutely horizontal, oil rings tend to "run downhill" and make contact with stationary components.

   (b) Ring movement will be erratic and ring edges will undergo abrasive wear. The oil will be seriously contaminated.

(c) The product of shaft diameter (inches) and shaft speed (rpm) should be kept below 8,000. Thus, a 3-inch shaft operating at 3,600 rpm (DN = 10,800) would not meet the low-risk criteria!

(d) Operation in lubricants that are either too viscous or not viscous enough will not give optimized ring performance and may jeopardize bearing life.

(e) The depth of immersion must be closely controlled, the bore finish must be 16 RMS or better, and the ring eccentricity should not be allowed to exceed 0.002 inches (0.05 mm).

- Flinger spools or flinger discs fastened to pump shafts often perform much more reliably than oil rings (slinger rings).
  (a) Consider retrofits using metal flingers discs; realize that cartridge-mounted bearings may be needed.
  (b) Some retrofits are made with metal hubs/cores to which elastomeric discs are firmly fused or otherwise attached. Be careful and apply these only within the manufacturer-approved peripheral speed range.

- If the use of oil rings is unavoidable, then be aware that a 30-degree angle between the contact point at the top of a shaft and points of entry into the oil represents the proper depth of immersion. Too much immersion depth will cause rings to slow, whereas insufficient depth tends to deprive bearings of lubricant.

- Oil rings with circumferentially machined grooves will provide increased oil flow. They still require all of the previous risk reduction steps.

Constant Level Lubricators

- Fully consider vulnerabilities of unbalanced constant level lubricators. If you must use constant level lubricators, then use only pressure-balanced models.

- Mount constant level lubricators on the correct side of the bearing housing. Observe "up-arrow" provided by the manufacturer of constant level lubricators. Incorrect mounting will lead to greater disturbances around the air-oil interface in the surge chamber of constant level lubricators. Mounting on the "up-arrow" side of the bearing housing reduces the height difference between uppermost and lowermost oil levels. In other words, it ensures a more limited level variation.

- Recognize that constant level lubricators are maintenance items that will require periodic replacement. The pliable caulking between a transparent bulb and its metal base will degrade over time, and small cracks will allow water to be pulled in by capillary action.

BEARING HOUSING PROTECTOR SEALS ("BEARING ISOLATORS")

- Consider buying only true state-of-art bearing housing seals to preclude ingress of atmospheric contaminants and egress of lubricating oil. Avoid old-style bearing isolators that
    (a) Use a dynamic O-ring in close proximity to the sharp edges of an O-ring groove (a damaged or nonfunctioning O-ring greatly increases the risk of black oil in the form of O-ring debris)
    (b) By using only a single clamping ring risk the rotor to skew or "walk" away from its intended location on the shaft
    (c) Are not field-repairable

- Install two large bulls-eye sight glasses on opposite sides of a pump bearing housing to view the actual operating oil levels.

MOTOR LUBRICATION SUMMARY

- Overgreasing of electric motor bearings is responsible for more bearing failures than grease deprivation. Know where the spent grease ends up—hopefully not in the motor windings. Practicing proper regreasing procedures is essential for long bearing life.

- Lifetime-lubricated (sealed) bearings will last only as long as enough grease remains in serviceable condition within the sealed cavity. Whenever the product of bearing bore (mm) multiplied by shaft rotational speed (rpm) exceeds 80,000, reliability-focused plants consider it uneconomical to use lifetime-lubricated bearings in continuously operating industrial machinery.

- Grease replenishing intervals depend on bearing inner ring bore dimension and shaft rotational speed. Reliability-focused user plants consider $dn = 300,000$ (d = bearing bore, mm; n = shaft rotational speed, rpm) the maximum for grease lubrication of electric motors and other machines in continuous service. It has been reasoned that, beyond this dn value, grease replenishing intervals become excessively frequent and oil lubrication would be more economical.

- On grease-lubricated couplings, verify that only approved coupling greases are used. Most motor bearing greases will centrifuge apart at high coupling peripheral speeds. Do not use an electric motor grease (EM grease) in a gear coupling.

- Conversely, do not allow coupling greases to be used in electric motor bearings. Most motor bearings will fail prematurely unless a premium grade "EM" grease is used.

- The advertised "all purpose" greases are not suitable for electric motor driver bearings in reliability-focused plants.

- Verify that relubrication and grease replenishment procedures take into account that:

  (a) Certain grease formulations cannot be mixed with other grease types. Mixing of incompatible greases will typically cause bearing failures within 1 year.
  (b) Attempted relubrication without removing grease drain plugs will often cause the grease cavity to be pressurized.
  (c) Overgreasing will cause excessive temperatures. On shielded bearings, cavity pressurization tends to push the shield into contact with rolling elements or bearing cage, causing extreme heat and wear.

MECHANICAL SEAL ISSUES

- It is generally acknowledged that most pump failure incidents involve mechanical seal distress. Although this is true at many facilities, it is also true that a major refinery has documented an average mechanical

seal life in excess of 10 years (Ref. 1). Using the right selection and installation procedures can markedly improve seal life and reduce pump
failure incidents.

- Select mechanical seal types, configurations, materials, balance ratios,
  pressure-velocity (p-v) values, and flush plans certified to represent
  proven experience in identical services or under verified-to-be-compa-
  rable service and operating conditions. Only they can guarantee suitable or extended seal life.

- Except for gas seals ("dry seals"), mechanical seals must be operated
  to preclude liquid vaporization between faces. However, using cooling
  water in a jacketed seal chamber cannot effectively cool the seal environment. External cooling of the flush liquid is far more effective.

- Mechanical seals with quench steam provisions are prone to fail rapidly
  if quench steam flow rates or pressures are not kept sufficiently low.
  Installing small diameter fixed orifices will limit excessive steam quench
  rates.

- Select the optimum seal housing geometry and dimensional envelope
  to improve seal life. Recognize that slurry pumps generally benefit from
  steeply tapered seal housing bores. The traditional concentrically bored
  stuffing box environment does not usually represent the optimum configuration for slurry pumps.

- Avoid inefficient pumping rings (the ones with small cogs machined
  into the periphery) on dual seals.

  (a) Axial pumping screw arrangements are more efficient, but for
      greater efficiency, some manufacturers use unacceptably close
      clearances (as little as 0.010 inches/0.25 mm) that violate the spirit
      of the 2000-vintage reliability-focused API specification clauses.
  (b) Axial pumping screws with close clearances risk making contact
      with the surrounding sleeve; this could touch off more massive seal
      failures.
  (c) Bidirectional tapered pumping devices for dual seals reduce failure
      probability; they have open clearances (typically ~0.060 inches/1.5

mm) and promote sealing or barrier fluid flow at optimized head and flow rate (H/Q) ratios.

- On hot service pumps, follow approved warm-up procedures. Verify that seal regions are exposed to through-flow of warm-up fluid (i.e., are not dead-ended).

- Understand the difference between conventional mechanical seals (seals in which the flexing portion rotates) and stationary seals (seals in which the flexing part is stationary).
  (a) Maximum allowable speeds and permissible shaft run-outs are lower for conventional seals and are higher for stationary seals.
  (b) Seals with springs exposed to the pumpage may be more prone to fail than seals with springs flooded by a more benign liquid or gas (air) environment.

- Flush plans routing liquid from the stuffing box back to suction generally require a pressure difference in excess of 25 psi (172 kPa).

- Fluid temperatures in a seal cavity must be low enough to prevent fluid vaporization in seal faces.

HYDRAULIC ISSUES

- Determine the suction energy and net positive suction heat (NPSH) margin (difference between NPSH available and NPSH required) for the application. If dealing with a "high" or "very high" suction energy pump (per HI definition), then make sure that

  (a) The pump is not operating in the suction recirculation region
  (b) Adequate NPSH margin has been provided
  (c) The installation has little or no pipe stress (i.e. good piping practices have been followed)

- Ascertain that pumps operating in parallel have closely matched operating points and share the load equally. Examine slopes of performance curves for each. Understand that differences in internal surface roughness may cause seemingly identical pumps to operate at different flow/pressure points.

(a) Although centrifugal pump life can be greatly curtailed when operating in the low-flow range where impeller-internal flow recirculation is likely to exist, "last resort" help may be of value. A concentric "flow tube" inserted into a spool piece adjacent to the pump suction nozzle will reduce recirculation severity.

(b) On pumps with power inputs greater than 230 kW, verify that "gap A," the radial distance between impeller disc tip and stationary parts, is in the range of only 0.050-0.060 inches (1.2-1.5 mm). Pump-internal recirculation is thus kept to a minimum. On reduced diameter impellers, this would imply that trimming is done only on vane tips! The impeller covers (shrouds) and discs remain at full diameters.

(c) On pumps with power inputs greater than 230 kW, ascertain that "gap B," the radial distance between impeller vane tips and cutwater, is somewhere in the range of 6% of the impeller radius. This will reduce vibratory amplitudes occurring at blade passing frequency.

(d) Ensure hydraulically induced shaft deflections in single volute pumps are not excessive. This may require restricting the allowable flow range to an area close to best efficiency point (BEP).

(e) Operate two-stage overhung pumps only at flows within 10% of BEP.

(f) Recognize severe shaft deflection and risk of shaft failure caused by reverse bending fatigue when operating far away from BEP.

(g) Check impeller specific speed versus efficiency at partial flow conditions. Consider installation of more suitable impellers for energy conservation.

(h) Consider installing in the existing pump casing an impeller with different width, a different impeller vane angle, a different number of vanes, or combinations of these. Observe resulting change trends in performance curves.

(i) Consider changing the slope of performance curve by inserting restriction bushing in the pump discharge nozzle.

(j) Review whether NPSH gain by cooling the pumpage is feasible and economically justified.

(k) Consider extending the allowable flow range by using an impeller with higher NPSHr. Verify that NPSHa exceeds the NPSHr of the new impeller.

(l) Use a ratio NPSHa/NPSHr of 3:1 or higher for carbamate and similar difficult services. There are many services in which several feet or meters of difference between NPSHa and NPSHr is not sufficient to prevent caviatation.

(m) Be aware of prerotation vortices and their NPSHr-raising effects on mixed flow pumps (i.e., pumps in certain specific speed ranges).

(n) Consider use of a vertical column pump or placing pump below grade if NPSHa-gain is needed.

(o) Consider inducer-type impellers where lower NPSHr is needed, but be aware that, to the right and left of BEP, the new NPSHr may now actually be higher than before!

(p) Mechanical improvement options are often related to specifications, work procedures, and mechanical workforce training. Again, there may be some overlap because every job function in a process plant has potential impact on equipment reliability. Note, also, that several items relate to component upgrading. In reliability-focused facilities, every repair event is viewed as an opportunity to consider upgrade options.

IMPELLER HYDRAULICS

(a) Consider the effects on a performance curve that could result from:
    (i) Vane underfiling

(ii) Vane overfiling

(iii) "Volute chipping"

(c) Calculate axial thrust values and verify adequacy of thrust disc or calculate balance piston geometry. Modify balance disc or balance piston diameter, as required.

(d) Consider opening existing impeller balance holes if axial thrust must be reduced to extend bearing life. Realize that vendor-supplied balance holes are not always correctly dimensioned and may have to be opened.

(e) Review and implement straight-run requirements for suction piping near pump inlet flanges entry. Aim for a straight pipe run of at least five pipe diameters between an elbow and the pump suction nozzle. Consult Ref. 1 or Hydraulic Institute (HI) guidelines for more precise recommendations.

(f) Realize that two elbows in suction piping at 90 degrees to each other tend to create swirling and prerotation. In this case, use ten pipe diameters of straight run piping between the pump suction nozzle and the next elbow.

(g) On top suction pumps, maintain a ten-pipe diameter straight pipe length between suction block valve and pump suction nozzle.

(h) Verify that eccentric reducers in suction lines are installed with flat side at the top to avoid air or vapor pockets. Note the exception in installations with pumped fluid entering suction nozzle from overhead location.

(i) Use pumping vanes or suitably dimensioned impeller balance holes to reduce axial load acting on thrust bearings.

(j) Oblique trimming of impeller vane exit tips, but retaining equal (Gap "A" compliant) cover (shroud) and disc diameters, will reduce the severity of vibratory amplitudes at blade passing frequencies and their harmonics.

(k) Oblique trimming of impeller vane exit tips, but retaining the full diameter only on the cover (suction side) will make the head-flow curve steeper while still developing maximum head.

(l) "Scalloping" of unsupported regions will improve internal flow profile while reducing risk of cracking and fatigue failure.

MECHANICAL IMPROVEMENT OR UPGRADE OPTIONS

• Implement suitable vortex breaker baffles on large vertical sump pumps.

(a) Implement wear ring modifications to reduce severity of rub in the event of contact because of excessive shaft deflection or run-out. Consider using high-performance graphite fiber polyimide resin for all wear rings or throat and throttle bushings.

(b) Examine the need for occasional measures to cure plate-mode or impeller cover (shroud) vibration. Consider "scalloping," if necessary to avoid impeller vibration other than unbalance-related vibratory action.

(d) Use generous fillet radii (0.2 inches, or 5 mm minimum) at shaft shoulders in contact with overhung impellers to avert reverse bending fatigue failures.

• If the calculated shaft deflection exceeds 0.002 inches (0.05 mm) at any of the anticipated flow conditions, then the shaft is probably too slender for reliable long-term operation. A superior replacement pump might be considered.

• Verify that shaft slenderness is not excessive. On old API-610/5th Edition pumps, the stabilizing effect of packing may have been lost when converting to mechanical seals. Therefore, throat bushings may have to be replaced by minimum clearance (0.003 inches/inch, or 0.003 mm/mm shaft diameter) shaft support bushings. These high-performance graphite fiber polyimide resin bushings should be wider than the customary open-clearance throat bushings originally installed.

(a) Verify absence of shaft critical speeds on vertical pumps. Insist on conservative bearing spacing.

(b) Verify acceptability of equipment spacing in pump pits as well as ascertain conservatism of sump design. View HI guidelines for spacing details.

(c) Consider hollow-shaft motor drivers on vertical pumps and always use suitable reverse rotation prevention assemblies.

(d) Beware exceeding the rule-of-thumb maximum allowable impeller diameter for 3,600 rpm overhung pumps: 15 inches (~380 mm).

(e) Consider in-between-bearing pump rotors whenever the product of power input and rotational speed (kW times rpm) exceeds 675,000.

(f) To survive or to reach long trouble-free operating times, centrifugal pumps must be properly installed. Installation checklists must be used, and accountabilities must be defined.

PROCESS PUMP REPAIR DIMENSIONS

All of the following refer to typical refinery pumps. These general guidelines may be used if the manufacturer's more specific instructions are no longer available.

01. Radial ball bearing inside diameter (I.D.) to shaft fits: 0.0001 inches-0.0007inches (2.5-17 μ) interference.

02. Radial ball bearing outside diameter (O.D.) to housing fits: 0.0001 inches-0.0015 inches (2.5-37 μ) clearance.

03. Back-to-back mounted thrust bearing I.D. to shaft fits: 0.0001 inches-0.0005 inches (2.5-13 μ) interference.

04. Back-to-back mounted thrust bearing O.D. to housing fits: 0.0001 inches-0.0015 inches (2.5-37 μ) clearance.

05. Shaft shoulders at bearing locations must be square with shaft centerlines within 0.0005 inches (13 μ).

06. Shaft shoulder height must be 65-75% of the height of the adjacent bearing inner ring.

07. Sleeve-to-shaft fits are to be kept within 0.001 inch-0.0015 inches (25-37 µ) clearance.

08. Impeller-to-shaft fits, on single-stage, overhung pumps are preferably 0.0000 inches-0.0005 inches (0-13 µ) clearance fits.

09. Impeller-to-shaft fits, on multistage pumps, sometimes require interference fits and must be checked against the manufacturer's or in-house reliability professionals' specifications.

10. Keys should be located in keyways with 0.0000 inch-0.0001 inch (0-2.5 µ) interference.

11. In view of point 10, keys should be hand-fitted to a "snug fit."

12. (a) Throat bushing-to-case fit is to be 0.002 inches-0.003 inches (50-75 µ) interference.
    (b) Throat bushing-to-shaft fit is to be 0.015 inches-0.020 inches (0.4-0.5 mm) clearance.
    (c) Throat bushing-to-shaft fits of inline pumps (depending on shaft size) where the throat bushings may act as intermediate bearings, will have clearances ranging from 0.003 inches to 0.012 inches (0.075-0.27 mm).

13. Weld overlays can be substituted for impeller wear rings. When separate wear rings are used, the wear ring-to-impeller fit should be 0.002 inches-0.003 inches (0.05-0.08 mm) interference.

14. Impeller wear rings should be secured by either dowelling, set screws that are threaded in the axial direction partly into the impeller and partly into the wear ring, or tack-welded in at least two places.

15. Clearance between impeller wear ring and case wear ring should be 0.010 inches to 0.012 inches plus 0.001 inches per inch up to a ring diameter of 12 inches. Add 0.0005 inches per inch of ring diameter over 12 inches.

16. For pumping temperatures of 500°F (260°C) and higher, add 0.010 inches (0.25 mm). Also, whenever galling-prone wear ring materials

(such as stainless steel) are used, 0.005 inches (0.13 mm) are added to the clearance.

17. Impeller wear rings should be replaced when the new clearance reaches twice the original value.

18. Case wear rings are not to be bored out larger than 3% of the original diameter.

19. Case ring-to-case interference should be 0.002 inches-0.003 inches (0.05-0.07 mm).

20. Case rings should be secured (doweled or spot welded) in two or three places.

21. Oil deflectors do not take the place of modern bearing protector seals. For best results and if still used, oil deflectors should be mounted with a shaft clearance of 0.002 inches-0.003 inches (0.05-0.07 mm).

22. On packed pumps, the packing gland typically has a shaft clearance of 0.030 inches (0.8 mm) and a stuffing box bore clearance of 0.015 inches (0.4 mm). On packed pumps, (a) the lantern ring clearance to the shaft is typically 0.015 inches-0.020 inches (0.4-0.5 mm) (b) the lantern ring clearance to the stuffing box bore is typically 0.005 inches-0.010 inches (0.13-0.25 mm).

23. Coupling-to-shaft fits are different for pumps below 400-hp (300 kW) driver rating and pumps with drivers of 400 hp and higher. Below 400-hp pumps employ fits metal-to-metal to about 0.0005 inches (~13 µ), whereas 400 hp and higher models often use interference fits ranging from 0.0005 inches-0.002 inches (13-50 µ).

    Taper bore coupling and hydraulic dilation fits should be as defined by either the manufacturer or your in-plant reliability professionals.

23. On pumps equipped with mechanical seals:
    (a) Seal gland throttle bushing to shaft clearance should typically be 0.018 inches-0.020 inches (0.4-0.5 mm), unless the pump is in high-temperature (HT) service.

(b)　For HT service applications (greater than 500°F/260°C), a somewhat larger clearance may be specified by the manufacturer or your plant reliability professionals.

(c)　Seal locking collar to shaft dimensions are typically 0.002 inches-0.004 inches (0.05-0.1 mm) clearance.

24.　Heads, suction covers, adapter pieces (if used), and bearing housing to case alignment should have 0.001 inches-0.004 inches (0.03-0.10 mm) clearance.

What We Have Learned

Checklists or other written reminders are of value to the pump repair person. They merit periodic updating and should be reviewed by personnel representing the three job functions that come into almost daily contact with process pumps: The operators, mechanical/maintenance work force members, and project/technical staff at virtually every facility.

Whenever the specific dimensional guidelines from the pump manufacturer are different from the generalized guidelines found in this text, the manufacturer's guidelines should be used.

If manufacturers' guidelines are no longer available, then the experience-based values and dimensions listed here can be implemented with confidence by the process pump user.

References

1.　Bloch, Heinz P., and Alan R. Budris; *Pump User's Handbook*, 3rd Edition (2010), The Fairmont Press, Inc., Lilburn, GA. (ISBN 0-88173-627-9).

2.　Bloch, Heinz P.; "Use laser-optics for machinery alignment," *Hydrocarbon Processing*, October 1987.

3.　Bloch, Heinz P.; "Laser-optisches Maschinenausrichten," *Antriebstechnik*, Volume 29, Number 1, June 1990.

4.　Bloch, Heinz P.; "Update your shaft alignment knowledge," *Pumps & Systems*, December, 2003.

5.　Bloch, Heinz P.; *Machinery Reliability Improvement*, (1982), Gulf Publishing Company, Houston, TX. Also revised 2nd & 3rd Editions. (ISBN 0-88415-663-3).

Chapter 16

Failure Statistics and Structured Failure Analysis

It is important to know how well a particular facility's pumps perform compared with those installed at a competitor's plants. There probably is room for improvement if a plant does not measure up to above-average pump failure statistics.

MEAN-TIME-BETWEEN FAILURES (MTBF) AND REPAIR COST CALCULATIONS

An easy comparison among pump users is feasible. It consists of adding up all process pumps installed at a plant and then to divide by the number of pump repairs per year. For a well-managed and reasonably reliability-focused U.S. refinery with 1,200 installed pumps and 156 repair incidents in 1 year, the mean time between failures (MTBF) would be (1,200/156) = 7.7 years.

The refinery would count as a repair incident the replacement of parts—any parts—regardless of cost. In this case, a drain plug worth $2.00 or an alloy impeller costing $5,000 would show up the same way on the MTBF statistics. Only the replacement or change of lube oil would not be counted as a repair.

A best-practices plant counts in its total pump repair cost all direct labor, materials, indirect labor and overhead, administration cost, and the cost of labor to procure parts. It assigns a value to failure avoidance, even the prorated value of avoiding pump-related fire incidents. Likewise, it assigns a monetary value to a work force relieved of pump repair burdens and assignable to proactive asset failure avoidance tasks.

Typical published pump repair costs have averaged $10,287 in 1984 and $12,000 in 2008. After inflation is factored in a repair, an actual cost reduction trend is indicated over this 24-year time span. Predictive maintenance and better pump monitoring may have contributed to reduced failure severity on the typical pump, although the ultimate consequences of some pump failures are grave.

The mean-times-between-failures (installed life before failure) of Table 16.1 have been estimated in 2004. Published data and observations made in the course of performing maintenance effectiveness studies and reliability audits in the late 1990s and early 2000s were used in these estimates.

Seal life statistics were estimated in the early 2000s (Table 16.2). These led to an estimation of reasonable goals (Table 16.3). Note that "target" is less than "best actually achieved."

Many plants achieve the months of installed lives indicated in Tables 16.2 and 16.3. Note that the actual operating life of a component would thus be about one-half of its installed life. To reach these pump life expectancies,

Table 16.1: Pump mean times between failures (Ref. 1).

• ANSI pumps, average, USA:	2.5 yrs / 30 mo
• ANSI/ISO pumps average, Scandinavian P&P plants:	3.5 yrs / 42 mo
• API pumps, average, USA:	5.5 yrs / 60 mo
• API pumps, average, Western Europe:	6.1 yrs / 73 mo
• API pumps, repair-focused refinery, developing country:	1.6 yrs / 19 mo
• API pumps, Caribbean region,	3.9 yrs / 47 mo
• API pumps, best-of-class, U.S. Refinery, California:	9.2 yrs / 110 mo
• All pumps, best-of-class petrochemical plant, USA (Texas):	10.1 yrs / 121 mo
• All pumps, major petrochemical company, USA (Texas):	7.5 yrs / 90 mo

Table 16.2: Suggested refinery seal target MTBFs (Ref. 1).

Target for seal MTBF in oil refineries	
Excellent	>90 months
Very good	70/90 months
Average	70 months
Fair	62/70 months
Poor	<62 months

Table 16-3: Component life targets (Ref. 1).

Seals		
	Refineries	*Chemical and other plants*
Excellent	90 months	55 months
Average	70 months	45 months

Couplings		
All plants	Membrane type	120 months
	Gear type	> 60 months

Bearings		
All plants	Continuous operation:	60 months
	spared operation	120 months

Pumps	
Based on series system calculation	48 months

the pump components themselves must be operating at the highest levels of reliability. An unsuitable seal with a lifetime of 1 month or less would have a serious negative effect on pump MTBF, as would an under-performing coupling or bearing.

Performing Your Own Projected MTBF Calculations

Simplified calculations will give an indication of the extent to which improving one or two key pump components can improve overall pump MTBF (Ref. 2).

Say, for example, that there is an agreement that the mechanical seal is the pump component with the shortest life followed by the bearings, coupling, shaft, and sometimes impeller, in that order. The anticipated mean-time-between-failure (operating MTBF) of a complete pump assembly can be approximated by summing the individual MTBF rates of the individual components, using the following expression:

$$1/MTBF = [(1/L_1)^2 + (1/L_2)^2 + (1/L_3)^2 + (1/L_4)^2]^{0.5} \quad \text{Eq. (16.1)}$$

In a 1980s study, the problem of mechanical seal life was investigated. An assessment was made of probable failure avoidance that would result if shaft deflections could be reduced. It was decided that limiting shaft deflection at the seal face to a maximum of 0.001 inch (0.025 mm) probably would increase seal life by 10%. It was thought that increasing seal housing dimensions to accommodate modern seal configurations would more than double seal MTBF.

All components that could be upgraded were examined. The life estimates were collected and then used in MTBF calculations.

In Equation (16.1), L_1, L_2, L_3, and L_4 represent the life, in years, of the component subject to failure. Using applicable data collected by a large petrochemical company in the 1980s, mean-times-between-failures and estimated values for a reliability-upgraded pump were calculated. The results are presented in Table 16.4. As an example, a standard construction ANSI B73.1 pump with a mechanical seal MTBF of 1.2, bearing MTBF of 3.0, coupling MTBF of 4.0, and shaft MTBF value of 15.0, resulted in a total pump MTBF of 1.07 years (actual operating hours). By upgrading the seal and bearings, the estimated achievable pump MTBF (actual operating hours) can be improved by 80%, to 1.93 years.

Table 16.4 shows the influence of selectively upgrading either bearings or seals or both on the overall pump MTBF. Choosing a 2.4-year MTBF seal and a 6-year MTBF bearing (easily achieved by preventing lube oil contamination via superior bearing housing protector seals) had a major impact on increasing the pump MTBF. Assuming the upgrade cost is reasonable, better seals are the best choice.

Based on year 2002 reports, a typical ANSI pump repair costs $5,000. This average cost includes material, parts, labor, and overhead. Assume that the MTBF for a particular pump is 12 months and that it could be extended to 18 months. This would result in a cost avoidance of $2,500/year—which is greater than the premium one would pay for the reliability-upgraded centrifugal pump.

Reduced power demand would, in many cases, improve the payback. Selecting advantageous pump hydraulics benefits both pump life and operating efficiency. Audits of two large U.S. plants identified seemingly small pump and pumping system efficiency gains that resulted in power-cost savings of many hundreds of thousands of dollars per year. Thus, the primary advantages of reliability-upgraded process pumps are extended operating life, higher operating efficiency, and lower operating and maintenance costs.

Table 16.4 provides a quick means of approximating the annual pump repair frequency based on the total (installed life) MTBF. Equation (16.1) and Table 16.4 also can be used to determine potential savings from upgrades and should shape the pump user's strategies.

An experience-based observation assumes that every missed upgrade item reduces pump life by 10% to a new life factor of only 0.9 years. If we miss six such upgrade items (and they are all discussed in the preceding chapters), then we will have reduced the anticipated life or MTBF to $0.9^6 = 54\%$ of what it might otherwise have been.

OLDER PUMPS VERSUS NEWER PUMPS

After 50 or 60 years of service and many maintenance actions, numerous "standard" American National Standards Institute (ANSI) and International Standards Institute (ISO)-compliant pumps are still operating. When they were designed in the 1950s and 1960s time frame, frequent repairs were accepted. Also, plant maintenance departments were staffed with more personnel than in later decades.

Unless selectively upgraded, a decades-old "standard" pump population will not allow 21st century facilities to reach their true reliability and profitability potentials. An older pump will generally fail more often than a newer pump. Likewise, a standard process pump will fail more often than an upgraded process pump.

Buried in Table 16.1 is a plant with more than 2,000 installed pumps; their average size is close to 30 hp. In 2010, this pump population had an

Table 16.4: How selective component upgrading influences MTBF.

ANSI Pump Upgrade Measure	Seal MTBF (yrs)	Bearing MTBF (yrs)	Coupling MTBF (yrs)	Shaft MTBF (yrs)	Composite Pump MTBF (yrs)
None, i.e. "Standard"	1.2	3.0	4.0	15.0	1.07
Seal and Bearings	2.4	6.0	4.0	15.0	1.93
Seal Housing Only	2.4	3.0	4.0	15.0	1.69
Bearing Environment	1.2	6.0	4.0	15.0	1.13

MTBF of slightly more than 9 years. Its owner-operators prided themselves in cultivating effective interaction between the mechanical and process-technical workforce members. The reliability professionals at this plant fully understood that pumps are part of a system and that the system must be correctly designed, installed, and operated if high reliability is to be achieved with consistency. It should also be pointed out that this plant (and others in its peer group) conducted periodic pump reliability reviews.

RELIABILITY REVIEWS START BEFORE PURCHASE

The best time for the first reliability review is before the time of purchase. This subject is given thorough treatment in Ref. 3. Individuals with reliability engineering background and an acute awareness of how and why pumps fail are best equipped to conduct such reviews. Trained reliability professionals should have an involvement in the initial pump selection process. Individually or as a team, those involved should consider the possible impact of several issues, including the ones mentioned earlier. All issues merit close attention and are again summarized for emphasis:

• Keep in mind the potential value of selecting pumps that might cost more initially but last much longer between repairs. The MTBF of a better pump may be 1-4 years longer than that of its nonupgraded counterpart.

• Consider that published average values of avoided pump failures range from $2,600 to more than $12,000. This does not include lost opportunity costs.

• One pump fire occurs per 1,000 failures. Fewer pump failures means fewer destructive pump fires.

Remember that there are several critically important applications in which buying pumps on price alone is almost certain to cause an inordinately larger number of costly failures. Spending time and effort on pre-purchase reliability reviews (Ref. 3) makes much economic sense, especially when dealing with the following:

• Applications with insufficient net positive suction head (NPSH) or low NPSH margin ratios (Ref. 4)

- High or very high suction energy services
- High specific-speed pumps (Ref. 5)
- Feed and product pumps without which the plant could not operate
- High-pressure and high-discharge energy pumps
- Vertical turbine-style deep-well pumps

Diligent reviews concentrate on typical problems encountered with centrifugal process pumps; an attempt is made to eliminate these problems before the pumps ever reach the field. Among the most important problems that the reviews seek to avoid are:

- Pumps not meeting stated efficiency

- Lack of dimensional interchangeability

- Problems with timely delivery because vendor's sales and/or coordination personnel are being reassigned or no longer work there

- Seal problems and compromises in seal materials, flush plans, flush supplies, and so on

- Casting voids (repair procedures, maximum allowable pressures, metallurgy, etc.)

- Lube application or bearing problems (Ref. 6)

- Alignment, lack of registration fit (rabbetted fit), base plate weaknesses, grout holes too small, base plates without mounting pads, and ignorance of the merits of using pre-grouted base plates

- Missing documentation or manuals and drawings shipped too late

- Pumps that will not perform well when operating away from the best efficiency point (i.e., pumps that risk experiencing internal recirculation [Ref. 7])

STRUCTURED FAILURE ANALYSIS STRATEGIES SOLVE PROBLEMS

Repeat pump failures are an indication that the root cause of a problem was not found. Alternatively, and if the problem cause is known, someone must have decided not to do anything about it. Pursuing a structured failure analysis approach is necessary to solve problems. Guessing or "going by feel" will never do.

Structured analysis means a repeatable approach that can be learned and employed by more than one person (Ref. 8). Once an accurate analysis is documented, remedial steps can be agreed on and can be implemented. Also, whenever it can be established that a pump at location "A" suffers more failures than an identical pump at location "B," we can be sure that an explanation exists. The explanation is found in deviations from best practices in one or more of the following seven cause categories:

- Faulty design
- Material defects
- Fabrication and/or processing (machining) errors
- Assembly or installation defects
- Off-design or unintended service conditions
- Maintenance deficiencies, including neglect/procedures
- Improper operation

Searching for additional cause categories will not add value because anything uncovered will, at best, be a subset of these seven. However, if one systematically concentrates on eliminating five or six of the seven categories in succession, then one will arrive at the category in which a deviation exists. That will make it possible to concentrate on understanding what led to the deviation.

The pump person must pay close attention to the underappreciated, generally nonglamorous "basics" and do so before opting for the often costly and sometimes unnecessary high-tech solution. Pumps obey the laws of physics and there is always a cause-and-effect relationship. It follows that even seemingly elusive and generally costly repeat problems can usually be eliminated without spending much money.

An integrated, comprehensive approach to failure analysis starts out by either describing the deviation, or by stating the problem. Next, such an approach encourages, or even mandates, careful observation and definition of failure modes. The approach should employ preexisting or developed-as-you-go checklists and troubleshooting tables (Ref. 1). The already existing checklists are supplied by pump manufacturers and can also be found in a large body of literature.

The "FRETT" Approach to Eradicating
Repeat Failures of Pumps

From observation and examination of a failed part, one identifies failure agent(s), realizing that there are only four possibilities (Ref. 8):

- Force
- Reactive environment
- Time
- Temperature

It is extremely important to accept the basic premise that components will only fail because of one, or perhaps a combination of several, of these four failure agents. We use the acronym "FRETT" to recall these four agents.

Because there are no failure agents beyond these four, the troubleshooter must remain fully focused on these four agents. To reemphasize by an example, a bearing can only fail if it has been subjected to a deviation (or deviations) in allowable force ("F") or has been exposed to a reactive environment ("RE"), has been in service beyond its design life ("T"), or was subjected to temperatures outside the permissible range ("T").

The need for knowledge must not be overlooked. For instance, bearings can fail (overheat) when they are too lightly loaded. They will then skid—a topic that was discussed earlier in this text. But there we go again; skidding is traceable to an inadequate force ("F") and will manifest itself as a temperature excursion ("T"). Two of the four agents "FRETT" are at work.

Each failure, and indeed each problem incident, is the effect of a causal event. In other words, for every effect there is a cause, or there is a reason for every failure. Here is an example:

[Man Injured]—because man fell
    [Man Fell]—because man slipped
        [Man Slipped]—because there was oil on the floor
            [Oil on Floor]—because a gasket leaked

By arriving at the word "gasket," the cause-and-effect chain is focused at the component level. Once we have narrowed issues down to the component level, we know that one or sometimes two troublesome or unexpected or overlooked "FRETT" contributors must now be found. In this case, a gasket leaked. A gasket is clearly a component. So,

[Gasket Leaked]—Must be a result of: force? reactive environment? time? temperature? We must check it out on the basis of data. Without data, we would be guessing, and guessing does not lead to repeatable results.

— Force: Too much—Why do we rule it in or rule it out?
      Not enough—Why do we rule it in or rule it out?
— Reactive environment: Wrong material selected for the medium transported in the pipe?—Why do we rule it in or rule it out?
— Time: Was the same gasket left in place for many years?—Why do we rule it in or rule it out?
— Temperature: Too high? Too low? Which one of these two (or perhaps why both) might be ruled in or can confidently be ruled out in a particular instance?

The pump person must take a similar approach with pumps and other machinery. For every effect, there is a cause; there is a reason for every failure, and we have to find it:

[Pump is down]—because the shaft broke

[Shaft broke]—our failure mode inventory was consulted; let us assume we found the surface has fretting damage. That is a deviation from the norm

[Surface damaged]—because the coupling hub was loose. That would explain the fretting damage

An analyst can now try to get to the root cause by remembering that all pump failure events fit in one or more of the seven cause categories listed above. If the coupling hub was found to be loose, then what cause categories are likely and which ones can we eliminate?

— Design error? Unlikely, because other couplings are designed the same way and we have verified that they are holding well.
— Material defects? No, because a thorough metallurgical exam checks OK.
— Fabrication error? No, because the hardness checked OK; dimensional correctness was verified and had been recorded upon installation, 3 years ago.
— Assembly/installation defect? Suppose we have no data and defer it for possible consideration later.

— Off-design or unintended service conditions? No; we rule it out.
— Maintenance deficiencies (neglect/procedures)? No, because no maintenance (PM) is required on a coupling hub.
— Improper operation? No, because we have ascertained operator activities were in accordance with our established standards.

At this stage, the analyst would get back to what needs to be investigated or requires follow-up examination. This might be a good time to start compiling:

(a) A checklist of possible assembly errors: From discussions with maintenance personnel, we might conclude that none applies in this instance.
(b) A checklist of possible installation errors:
  • Force:—Could have overstretched hub.
    —Could have had insufficient axial advance on taper (insufficient interference fit).
  • Reactive environment: None found; normal chemical plant location.
  • Time: Ascertained that run length was not excessive; the hub failed after just a few weeks of operation.
  • Temperature: Suppose the coupling was heated to facilitate installation. How was the heat applied? What tells us that the temperature was within limits? The temperature could have been too high (causing overstretch) or too low (not allowing dilation to result in sufficient axial advance).

In both of these examples, the pump failure analyst has to determine in which cause category there is a deviation from the norm, which item needs to be modified, and how this modification must be implemented to prevent a repeat failure. Data will be required to support any conclusions. With data, one can define the root causes of a problem. Without data, one can, at best, determine a probable cause.

Change analysis parallels and supplements the structured, comprehensive approach. Change analysis seeks to identify what is different in the defective item as compared with an identical but unaffected item. The analyst probes into when, where, and why the change occurred. The analyst then outlines several remedial action steps and will have to choose the steps that best meet defined objectives. These objectives must achieve highest safety, and the analyst may pick from a list that includes lowest life-cycle cost, present value, highest initial quality, meeting a certain industry standard, a deadline, and so on.

The objective of aiming for lowest life-cycle cost usually makes considerable sense. Calculating this parameter would include the cost of staffing a pump selection or reliability review with dedicated, knowledgeable individuals. Life-cycle cost analyses must also include the value of downtime avoidance and MTBF extensions, as well as the value of avoided fire and safety incidents.

Recall that fewer pump failures translate into fewer fires and decreased insurance premiums. Failure avoidance creates goodwill and enhances a company's reputation. Also, having to cope with fewer failures frees up personnel whose proactive activities avoid other failures, and so on.

WHAT WE HAVE LEARNED

Needless to say, any choice we make will have its advantages and disadvantages. When pumps and process pump applications are involved, the most elementary choice requires opting for two out of three broad-brush deliverables: good, fast, and cheap. Take any two, but do not expect ever to obtain all three.

Whenever we are confronted with the two-out-of-three choice, we should remember that, for an analysis or repair to be good and fast, it probably will not be cheap. If we want it to be good and cheap, then it probably will not be fast. And if we opt to pursue the fast and cheap paths, it probably will not be good. In case we are persuaded to go the fast and cheap route, let us brace ourselves for repeat failures that can cost a small fortune and bring on all kinds of other misery.

Over the decades, we have come to realize that pump failure statistics are rarely scientific. Still, they are experience-based and should not be disregarded. If your MTBF hovers around average, then identify the repeat offenders and subject them to an uncompromising improvement program. In the hydrocarbon processing industry, about 7% of the pump population consumes 60% of the money spent on pump maintenance and repair. Getting at the root causes of failures on these 7% will save much money.

A strategy that involves rational thinking is solidly supported by a minute's worth of looking up vendor documentation. A sound strategy also mandates respect for the simple laws of physics. It is a strategy that results in failure cause identification; it will lead to future failure avoidance and will extend pump MTBF.

It can be said that all successful and cost-effective failure analysis methods represent structured approaches that give focus to an otherwise scattered search for the causes of equipment failures. Structured analysis approaches are repeatable; they aren't hit-or-miss guesses. A successful approach guides the user/analyst through a sequence of steps; it invariably accepts the premise that all problems are ultimately caused by the decisions, actions, inactions, omissions, or commissions of human beings. A successful approach is objective; it seeks explanations but does not tolerate compromises and excuses.

It is fitting, then, to conclude or recap this text by pointing to a simple illustration (Figure 16.1). This illustration tries to convey that many parameters interact to cause repeat failures in pumps. Many of these are classified as hydraulic issues, and much work has been done to improve pump hydraulics. However, the majority of what we chose to call elusive failure causes is linked to mechanical issues. We have become accustomed to maintenance routines that rarely question the adequacy of a vendor's design. Failure causes have become elusive because we overlook or forget (and even disregard) the laws of physics.

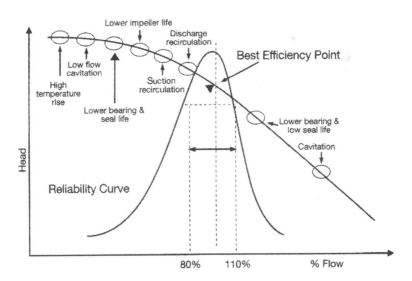

Figure 16.1: Staying near the center of this "Reliability Curve" is a wise course of action (Ref. 9)

In this text, we have also attempted to point out that process pump vendors often merely furnish a barely adequate design. Vendors are left with the impression that users are unwilling to pay for a superior design. Moreover, vendors and pump manufacturers benefit from the sale of replacement parts and are in business to generate income.

We must also not forget that pump manufacturers have right-sized, down-sized, and economized the way they do business. Few (if any) of these organizational realignments benefit the user and a preponderance of repeat failures attests to it. Some vendors and manufacturers no longer employ process pump experts and diligent craftsmen. The user-purchaser may belatedly come to realize that he has become the manufacturer's quality control inspector. Many must experience failures before they accept this fact. When they learn the hard way, they must allocate money to ward off this eventuality by suitable predelivery inspections.

Timely and competent up-front action by the owner-purchaser is one of the keys to failure avoidance. This up-front action includes development of detailed specifications for process pumps and some key components that go into good process pumps. Once a process pump arrives in the field, it must be properly installed and maintained. To be effective, the facility must adopt work processes and procedures that harmonize with best-of-class thinking.

To avoid repeat failures, pump owner-operators must deliberately push certain routine maintenance actions into the superior maintenance category. Superior maintenance efforts will lead to (or are synonymous with) pump reliability upgrading.

In essence, the course of wisdom demands that we move away from "business as usual." Before one can apply practical wisdom, one must acquire knowledge and understanding. We hope that this text has helped the reader in this regard.

References

1. Bloch, Heinz P., and Allan R. Budris; *Pump User's Handbook*, Second Edition (2006), The Fairmont Press, Lilburn, GA. (ISBN 0-88173-517-5).
2. Bloch, Heinz P., and Don Johnson; "Downtime Prompts Upgrading of Centrifugal Pumps," *Chemical Engineering*, November 25, 1985.
3. Bloch, Heinz P.; *Improving Machinery Reliability*, 3rd Edition (1998), Gulf Publishing Company, Houston, TX.
4. Karassik, Igor J.; "So, you are short on NPSH?" presented at Pacific Energy Association Pump Workshop, Ventura, CA, March 1979.
5. Ingram, James H.; "Pump reliability—where do you start," presented at ASME Petro-

leum Mechanical Engineering Workshop and Conference, Dallas, TX, September 13-15, 1981.

6. Bloch, Heinz P.; "Optimized lubrication of antifriction bearing for centrifugal pumps," ASLE, paper no. 78-AM-1D-1, presented in Dearborn, MI, April 17, 1978.

7. McQueen, Richard; "Minimum Flow Requirements for Centrifugal Pumps," *Pump World*, 1980, Volume 6, Number 2, pp. 10-15.

8. Bloch, Heinz P., and Fred K. Geitner; *Machinery Failure Analysis and Troubleshooting*, 3rd Edition (1997), Butterworth-Heinemann Publishing Company, Woburn, MA. (ISBN 0-88415-662-1).

9. Barringer, Paul; "API pump curve practices and effects on pump life from variability about BEP," Weibull Analysis Course. (See also Ref. 2, p. 621.)

Index

Printed in Poland
by Amazon Fulfillment
Poland Sp. z o.o., Wrocław
30 November 2020